CROSSRAIL
THE WHOLE STORY

CHRISTIAN WOLMAR is an author and broadcaster specialising in transport matters. He writes regularly for *The Times*, the *Guardian* and various other publications, and produces a fortnightly column for *Rail* magazine. He also appears frequently on TV and radio as a commentator and in documentaries.

CHRISTIAN WOLMAR

CROSSRAIL
THE WHOLE
STORY

HEAD of ZEUS

An Apollo Book

First published in the UK in 2018 by Head of Zeus Ltd
This paperback edition first published in the UK in 2022 by Head of Zeus Ltd,
part of Bloomsbury Publishing Plc

9 7 5 3 1 2 4 6 8

A catalogue record for this book is available from
the British Library.

ISBN (PB): 9781803281247
ISBN (E): 9781788540247

Typeset by Adrian McLaughlin

Printed and bound in Great Britain by
CPI Group (UK) Ltd, Croydon CRO 4YY

Head of Zeus Ltd
First Floor East
5–8 Hardwick Street
London EC1R 4RG

WWW.HEADOFZEUS.COM

Contents

Dedicated to my grandchildren and step-grandchildren, all boys, who seem, so far, to love trains – Alfie, Luka, Quinn and Louie. May they be travelling on Crossrail in the twenty-second century!

Preface

Writing this book has been a labour of love. I might have been sceptical of the Crossrail concept when I first wrote about it almost thirty years ago and the story was exclusively about money and politics. It seemed a rather banal concept, another tunnel under London like the Tube lines which have been part of my life, as I am a Londoner, since my childhood.

However, all the difficulties and faults with the project cannot take away from the fact that Crossrail will be a railway that Londoners will undoubtedly learn to admire and even to love. It is everything that a modern railway should be and its grandeur will put it in the same category as the Moscow Metro or the great stations built during the height of the railway age in cities across the world. I hope my book conveys this sense of excitement and achievement.

1.

The first Crossrail

Travelling between east and west London has always been difficult. While north–south journeys became much easier in the nineteenth century as more bridges were built and a couple of Underground lines crossed the river, the possibilities for east–west travel have always been limited because the sinuous path of the River Thames limits the potential for direct routes that avoid river crossings. Going along the embankment necessitates travelling a much greater distance because of the shape of the river and therefore north of the Thames there is only one direct alignment, the old A40, Shepherd's Bush to Lancaster Gate. Here, there is a spur along Sussex Gardens towards Marylebone Road, which offers a more circuitous route to the City, while the main route continues along Oxford Street and High Holborn to Bank. From the east, the two main routes, Commercial and Whitechapel Roads (the A11 and the A13), meet at Aldgate, where chaos ensues for those wishing to push further west.

As a result of this geography, the road into London from the west, the Roman Via Tribantia, was the most lucrative of the old turnpikes. In the mid-eighteenth century the tollgate at Tyburn – now Marble Arch – charged 10d* for a carriage with two horses, 4d for a horseman and 2d for twenty pigs. A toll at Notting Hill charged 4d for anyone who used it. For eastbound travellers entering London there was effectively no alternative to using these gates – and this remained largely the case throughout the first half of the nineteenth century.

The expansion of all modes of transport across London in the nineteenth century, including the revolutionary concept of building underground railways, was made necessary by the scale of London's growth during this period. From a relatively well-contained space encompassing much of what is Zone 1 of its transport system today (basically the area inside the Circle Line), London became the world's first megalopolis thanks to its place at the centre of a burgeoning colonial empire from which it gained extensive benefits. On its east–west axis, unconstrained by the barrier of a river, London in 1801 stretched for five miles and had a population of just under a million. By the end of the century it was seventeen miles wide with a population of more than 6.5 million. It was by far the biggest city in the world and its position at the heart of the empire ensured it became a magnet for the affluent, the ambitious and the footloose.

This extraordinarily rapid expansion stimulated the establishment of successive new transport systems to accommodate the growing number of people prepared to settle further and further outside the centre in order to find decent housing at affordable prices: new roads, omnibuses, trams (initially horse-drawn and later electric), the Metropolitan Railway and then, by the end

---

* There are 2.4 old pennies to a new penny.

of the century, deep Tube lines. Thanks to the constraints of the road system, it was on the east–west axis that many of these nineteenth-century transport initiatives were first introduced. It is not surprising, therefore, that the very paths they followed correspond to parts of the Crossrail route and the places they served are now the site of several of the line's stations.

The omnibus, London's first genuine public transport system, made its debut, as would the Metropolitan Railway a generation later, between Paddington and the City of London. The inaugural omnibus service, operated by George Shillibeer, 'an enterprising Bloomsbury mourning-coach builder with business connections in Paris'[1] (where such services were already well established), ran on that route in July 1829. His omnibuses, drawn by two horses, carried twenty-two people, thirteen of whom could be accommodated inside the stagecoach, protected from the elements, with the rest seated on top, where fares were lower. At around 6d even for the top deck, these fares were prohibitive for most Londoners.

The omnibus service was initially aimed at the capital's middle classes, but the need for faster transport developed so rapidly that within a few years omnibuses were serving all the major thoroughfares in London and carrying an impressive 200,000 passengers daily. These buses were the genesis of London's mass-transport system, but the inefficiency and high cost of horse-drawn transport, as well as the increasing congestion on the capital's main arteries, ensured that omnibuses could never, alone, meet the transport needs of the capital's growing population. In the mid-nineteenth century, as London filled up and expanded dramatically north of the river, none of the forms of transport in the capital was particularly enticing.

Although the railways soon expanded outwards from the large termini built on all four compass points, there was little effort to provide for the onward journeys of the increasing number of passengers arriving in London or for access to these new stations from other parts of the capital. Apart from walking, still the only option for most people including many in work, and the omnibuses, there were Hackney cabs, the horse-drawn forerunners of taxis, which, at 8d per mile were expensive and consequently out of reach of most of the population. London was therefore in need of a transport revolution.

The success of the world's first major railway line, the Liverpool & Manchester, which opened in 1830, had stimulated the remarkably rapid development of a network which, within just twenty years, encompassed 5,000 miles of track. London's first railway was on an east–west axis, but – perhaps surprisingly, given the pace and extent of the city's northern expansion – was in south London. Unlike most of the early lines, it was a service specifically designed as a suburban railway for commuters. The construction of the 3.5-mile London & Greenwich Railway in the late 1830s was a remarkable achievement since it was built on top of 878 arches which cut a swathe through southeast London and which to this day carry the large number of suburban services operating out of London Bridge station. On the whole, however, London's railways catered poorly for the capital's suburbs and the outlying villages that would soon be absorbed into the metropolis. This showed a lack of imagination on the part of those who were promoting London's railway lines: because they believed the role of the railways was to straddle the country and carry people long distances, they failed to realize that there was a potentially lucrative market on the capital's doorstep. As a result, the Great Northern Railway had just four stations in the eighteen miles between its grand terminus at King's Cross and the first significant town, Hatfield. The first stop out of Paddington on the Great Western Railway was, initially, West Drayton, near the site of today's Heathrow Airport. Even today there is a paucity of suburban services in west London, a factor that has caused huge problems for Crossrail's development and, indeed, as we shall see, will remain an issue for its operational effectiveness.

The other early London railway on the east–west axis was the North London Railway, which is described by the chronicler of its rather chequered history as having 'precious little to do with north London or the needs of its citizens at all'.[2] This is rather harsh but, in truth, reflects the fact that its original purpose was to provide access from the west to the economically important London docks and markets rather than serve local passengers. The railway was built in bits and pieces, and has had various incarnations, as well as the threat of closure, as recommended in the Beeching report of the early 1960s, hanging over it for several years. However, now it is an incredibly busy link between various north and northwestern suburbs and the eastern side of the City. The first section opened in 1850 between Bow Junction and Islington, and was soon extended to Camden and a junction with the London & North Western Railway, which owned the line (now the West Coast Main Line) in the west, and to Poplar Dock in the east. Thanks to a connection made a couple of years later at Willesden it stretched out to Kew and later Richmond, and when a new terminus at Broad Street, next to Liverpool Street, was built in 1865 it provided a route – albeit a very circuitous one – from London's western suburbs to the

City. Although the line's principal purpose was to carry freight, for a time in the nineteenth century it did become a significant passenger railway as the suburbs through which it passed grew rapidly. Indeed, its popularity led to so much congestion on the London & North Western Railway that in 1860 a new line was built through to Willesden Junction, including a three-quarter-mile tunnel under Hampstead, demonstrating its parent company's confidence in its viability.

However, the most significant railway development during this period was the far more forward-looking concept of building underground lines through the capital. The world's first, the Metropolitan Railway, owed its existence to Charles Pearson, one of those wonderfully eclectic Victorian characters whose legacy remains influential to this day. Pearson was a solicitor working for the City Corporation and had taken on a wide variety of social issues such as prison reform and discrimination against Jews and Catholics. He first came up with the idea of an underground railway that would run down the Fleet valley in a pamphlet published in 1845, and through his role in the City was able to mastermind the financing of the Metropolitan Railway, with construction eventually beginning in 1860.

The Metropolitan, described by Gillian Tindall in her book *The Tunnel Through Time* (2016) as 'Crossrail's first and most momentous predecessor', began at Bishop's Bridge, near the current Paddington station. It followed the line of the New Road, the turnpike built in the eighteenth century that was for a long time effectively London's northern boundary, later renamed Marylebone Road in the west and Euston Road in the east. The alignment under the road was chosen to avoid unnecessary and expensive demolition as the railway was built using the 'cut and cover' method. This involved digging a trench, fitting the railway and covering it over. The railway was only a few feet below ground which is why today's Metropolitan, District and Circle are called 'sub-surface' lines as opposed to the Tube lines bored later. The route was designed to ensure that the line served Euston and King's Cross stations (as well as, later, St Pancras, and, rather unsatisfactorily, Marylebone). It then bent south (with the alignment of the River Fleet) at King's Cross (or Battle Bridge as it had been known) along the Fleet valley to Farringdon, its original terminus. The line effectively followed the original route linking Roman Watling Street with the City of London. After a remarkably short construction period of just over three years, the Metropolitan Railway opened to much fanfare in January 1863. It was an immediate success.

The tollgates both along the New Road and at Marble Arch soon disappeared – to general rejoicing from local residents, one of whom was reported as saying: 'I regard the keeper of the tollgate as a legalised highwayman, he lays hold of your bridle, pats your horse and puts on the stance of a regular brigand telling you to stand and deliver.'[3] They were effectively made redundant by the construction of the Metropolitan even though the trains did not cater for horsemen let alone pigs.

The financial success of the initial section of the Metropolitan Railway between Paddington and Farringdon, built at the very reasonable cost of £1m, demonstrated the potential of sub-surface lines in cities. Since the early services were well used, the Metropolitan was soon extended westwards through to Kensington and a series of branches out to Ealing, Putney and Wimbledon were added to the system by the 1880s, giving many of the villages that are now part of west London an excellent service into town and ensuring that they remained villages no longer. Everywhere the railway went the developers soon followed. Hammersmith, a place which, thanks to its abundant water supply, produced excellent strawberries and spinach, was soon connected by two different lines and its fertile fields were rapidly turned over to housing. Within a couple of years of its opening, the railway was extended from Farringdon to Moorgate, precisely on the route that would be taken, some 150 years later and 25 metres (82 ft) deeper, by Crossrail.

Trams, which were hauled by horses until the last few years of the nineteenth century when electric power took over, were initially barred by Parliament from running in London and it was not until 1870 that legislation was passed to allow construction of four routes. The only one of these planned for the east–west axis, Kensington to Oxford Street, was never built and hostility to trams by the richer vestries (the forerunners of the boroughs that were formed at the end of the century) prevented any being built in a large swathe of west and central London. Consequently, tram services from the west only reached as far as Shepherd's Bush, which meant that Holland Park Avenue and Bayswater Road, the two roads linking the Bush with Marble Arch, never had a tram even though they were busy enough to have sustained a line.

One of the early tram experiments was an attempt in 1861, by the great and aptly named tram pioneer George Train, to lay down tracks for horse trams between Marble Arch and Porchester Terrace in Bayswater. However, objections from the horse-omnibus owners and – probably more significantly – the affluent local residents, who expressed concern that their horses would trip over the rails, forced him to rip up the track.

Oddly, more than a century and a half later, an attempt to build a tram line along the Uxbridge Road deep into outer west London was thwarted by huge public opposition and abandoned in 2007.

Instead, in late Victorian London the horse omnibuses did a roaring trade along the key thoroughfare from the west, which took in the Uxbridge and Bayswater roads as it was one of the busiest routes in London for these buses, and they remained profitable until the construction of a new Tube line, the Central Railway, at the turn of the century. The Central, in fact, only emerged after the failure of numerous other schemes to exploit what was plainly fertile ground for public transport. There were several attempts, starting in 1865, to build a railway along London's great east–west artery, which east of Marble Arch encompassed Oxford Street, High Holborn, Cheapside and Poultry, and these failed projects all had certain parallels with Crossrail in terms of their ambition, their importance to the capital and the financial, political and technical difficulties in getting them built.

No surface railway could be built into the centre of London because of a ruling by the Royal Commission on Metropolitan Railway Termini which in 1845 decreed that they would require too much demolition and cause too much disruption. Consequently London's stations were kept outside the existing built-up area, which is why all the major termini today are situated on what became a ring around the centre of the capital. The Commission's recommendation therefore necessitated the construction of the first underground lines that linked the termini with each other and the city centre. They were all built using the cut and cover method which caused much disruption on the surface and therefore it took until 1884 for the Circle Line,* which linked all London's major termini, to be completed.

The Commission's decision did not deter various entrepreneurs from making various proposals for new railways along the east–west axis. An Oxford Street & City Railway was proposed as early as 1865 but the following year succumbed to one of the Victorian era's regular financial crises. There were three more unsuccessful proposals for new lines in the next couple of decades. One, which would have gone from Marble Arch to the General Post Office near St Paul's, was to have been worked by nine stationary engines; another suggested a line under the city streets from Marble Arch to Whitechapel; while a third, in 1884, which went under the name of Central Railway, sought to run under Oxford Street, turning south at Oxford Circus to reach Trafalgar Square. All of these, like so many Victorian projects put forward by over-optimistic promoters, foundered because of lack of finance.

Two technological developments were to prove crucial in making such a project more feasible. The world's first Tube railway, the City & South London, opened in November 1890 showing that it was possible to bore London tunnels under a city without causing the buildings above to collapse. The key innovation was a device known as the Greathead Shield,† which enabled a team of workers to carve out tunnels while being protected from collapse, and allowed them to make far speedier

---

* A name it did not acquire until 1949 when it first appeared on the Tube map as a separate entity.

† Named after its inventor, James Henry Greathead (1844–96), the civil engineer responsible for the digging, in 1869, of the Tower Subway beneath the Thames, using the shield that bears his name.

progress than would otherwise have been possible. The shield is a cylindrical structure that protects the miners while they dig out the soil and the lining of the tunnel is installed, and it is pushed forward using a hydraulic system.

The second development was that the trains, which ran for just over three miles between Stockwell, south of the river, and King William Street, near Bank in the City of London, were powered by electricity rather than, as originally envisaged, cable. Given the number of bends and the sheer length of the line, cable would have been an unreliable and cumbersome method of traction. Electricity, despite being a relatively new source of energy, could nevertheless provide a more reliable and efficient service. There were hiccups, though. At times, the trains approaching the King William terminus, at the other end of the line from their power source at Stockwell, were unable to make it up the slight incline to the buffer stops and the driver had to allow the train to roll backwards before having a second go at climbing up to the end of the line. Not something that would happen today! Although the stations were gloomy because they were lit by gas in order to preserve the limited electricity supply for powering the trains, and the carriages were not fitted with windows because the line's promoters thought this was an unnecessary expense since there was nothing to see, the service was a great success. People flocked to use it, braving the descent into stygian darkness in a rattling lift and the subterranean journey in what became known as 'sardine cans'.

Emboldened by the construction of this line, a different Central London Railway company tried its luck in 1889, putting forward a parliamentary bill for a tube railway between Queens Road (now Queensway) in Bayswater to link up with the City

& South London at King William Street. However, the vagaries of the parliamentary process resulted in the proposal being rejected, a decision the influential magazine *The Railway Times* argued was the result of 'faddism aided by personal interest'.[4]

The motto of many of these intrepid Victorians, however, appears to have been 'if you fail, try again', and when the promoters of the Central Railway tried again the following year with a slightly longer project, they succeeded. The Act for the railway, passed in 1891, gave them permission to build a railway between Shepherd's Bush and Cornhill (soon extended to Bank and later to Liverpool Street), to be worked by electricity and to run in two tunnels of just 11 ft 6 in (3.5 metres) in diameter. This might have seemed generous at the time, but it soon proved to be inadequate for meeting technical requirements and, more importantly, London's transport needs. Crossrail's tunnels, in contrast, have a diameter of 6.2 metres (20 ft 4 in) – or, to put it another way, an area three-quarters greater in size.

Nevertheless, the Central Railway could be considered as the Crossrail of its day in terms of the ambition of the project. As ever, the promoters of the scheme, entirely dependent on private finance, struggled to raise the money. The promoters initially estimated that some £3.24m would be needed to build the line and in 1894 they tried to raise the capital on the stock market, promising that the line would be completed by the end of 1898. However, the issue was a failure, with less than half a million pounds being raised. This was hardly surprising, in that, despite the success of the City & South London, the numerous tube schemes promoted at the time were seen as a risky business. It was still relatively new technology and success was uncertain. So the promoters resorted to the oft-used Plan B of the time, which was to get a contractor or financier to fund the construction. In this case, the promoters, the Electric Traction Company, persuaded the banker and philanthropist Sir Ernest Cassel to fund most of the construction of this nineteenth-century Crossrail, with shareholders eventually chipping in the rest. Inevitably, there were delays and construction did not start until April 1896, when a shaft was sunk at Chancery Lane, at the very centre of the route. While the time taken to build the line exceeded the original estimate, work proceeded remarkably quickly despite the primitive nature of the excavation techniques available at the time. James Greathead died during the construction, but the shield to which he gave his name proved invaluable. Following his death, work on the line was overseen by Basil Mott, one of his colleagues on the City & South London, but fortunately his technique was relatively well established by then. In the rather chaotic private sector-led planning process of the day, disputes with other transport providers were commonplace and inevitably an argument with the Great Eastern and North London railways about the precise arrangements for the connection between the new tube line and their respective termini at Liverpool Street and neighbouring Broad Street developed into a stand-off. The resulting delay had to be sanctioned by an act of Parliament, but this was a relatively simple process at the time, since both MPs and Lords were eager to embrace the remarkable new railway.

One complication not suffered by modern-day Crossrail was the need for the alignment to be kept for the most part under streets rather than buildings, for fear of causing subsidence or even collapse. Consequently at three stations – Post Office (now St Paul's), Chancery Lane and Notting Hill Gate – the two tubes were built directly on top of each other in order to limit the width of the necessary bores and thereby to avoid having to dig out the earth underneath adjacent buildings.

The first trial trip along the whole line was made on 1 March 1900 and the Prince of Wales, soon to be Edward VII and a keen advocate for new technology, officially opened the line on 27 June at a ceremony attended by Mark Twain.* The Prince's non-stop journey between Bank – the section to Liverpool Street did not open until 1912 – and Shepherd's Bush took just eighteen minutes for the six-mile trip, undoubtedly far faster than would ever have been possible in the crowded streets above. It was a much grander enterprise than its predecessor, which had been built on the cheap and whose trains were now seen as outdated. The Central's huge electric locomotives, built in the camel-back style with the driver in the middle between two sloping engine compartments, each pulled half a dozen carriages, with large windows and gated entrances.

The *Financial Times* could not hide its enthusiasm: 'the new railway, so happily named the Central London, is undoubtedly the most important enterprise of its kind, not only in London or Great Britain, but in the world.'[5] This assessment was only a slight exaggeration. There were, by then, many impressive railways across the globe, but in running under a city's streets in tunnels carved out of the earth the Central had few rivals to the title of the world's most impressive *urban* line.

The opening of the railway may have been late according to the specification of the initial schedule, but the project was completed remarkably fast. From the sinking of the first shaft to the opening ceremony took just over four years, an astonishing

---

* Born Samuel Langhorne Clemens.

achievement considering it involved excavating twelve miles of running tunnels, creating eleven stations and installing seventeen signal boxes. Best of all, every station boasted electrically powered lifts, each capable of carrying 100 people and operated by a gateman, to take people down to the platforms which were far more spacious than those installed on the City & South London. The forty-nine lifts cost £90,000 and were all installed in time for the opening apart from those at Bond Street. The lifts proved amazingly resilient: the last surviving one, at Holland Park station, was replaced as recently as 1959. (I have a personal memory of that lift as I lived nearby as a boy and went each fortnight to watch Queens Park Rangers at Loftus Road, which involved hopping a couple of stops on the Central Line to White City.) These were not quite the first electric lifts on the system, since the City & South London had by then installed a handful. Hydraulic lifts, slower and more cumbersome, remained the norm. The size, efficiency and speed of these electric lifts greatly impressed the travelling public, helping to boost the number of early users of the line.

Given the pioneering nature of the enterprise, it was not surprising that a major problem quickly emerged after services started operating in earnest. The electric locomotives, which had been built by General Electric in the USA, were typical examples of heavy American engineering and proved unsuitable for Tube use. They weighed 44 tons and the powerful vibrations and noise they caused in adjoining buildings led to widespread complaints. The original idea had been to have two less heavy locomotives, one at the front and one at the back, to avoid the engine having to shunt around at the termini, but this was vetoed by the safety authority, the Board of Trade, because it would have involved connecting the two with a huge high-tension cable that was thought to be a fire risk.

Local householders were terrified that a building would collapse as a result of what must have seemed like a constant series of mini earthquakes. The Board of Trade responded by appointing a committee of three experts to assess the extent of the problem. They found that, as the trains passed, the draughtsmen in architects' offices in Cheapside were unable to draw straight lines, such was the vibration.

After an attempt to modify the locomotives failed, the Central Railway company decided to replace all the rolling stock with what are known as 'electric multiple units', which have motors underneath the carriages and therefore the weight is spread more evenly. This meant their rather elegant coaches had to be scrapped, but – in another testimony to how rapidly far-reaching changes could be made so efficiently at the time – the complete changeover to the new rolling stock took place in just two months, from the start of their introduction in April 1903. The draughtsmen and signwriters whose jobs had become almost impossible could now carry on in relative peace, and their landlords could rest easy about the risk of collapse as the new, much lighter, rolling stock produced little vibration.

There was, however, another problem – relating to the ventilation, or, rather, the lack of it – which would take longer to solve. The original idea was that the trains would push the air through a piston effect in the tight tunnels, which, it was expected, would create sufficient ventilation. It was a vain hope. The same air was simply being shunted around a virtually closed system, and it accumulated more dust and particulates from the movement of the trains. When chemists from the London County Council were called in to investigate, they discovered that the already fetid atmosphere was made even harder to breathe by the dryness of the air – which had an average humidity of 45 per cent, compared with a normal street level of 76 per cent – as well as by an excess of sulphur and nitrogen oxides. A succession of ad hoc solutions was partly successful in ameliorating, but not resolving, the situation. Fans were installed at various locations, but it was not until filtered and ozonized air was pumped into the system – a process started in 1911 but not completed until the 1930s – that the atmosphere began to improve. It was only after the Second World War that the present system, in which air is pushed around the line by a series of 130 fans, designed to keep the temperature at a rather warm 73°F (23°C), that standards acceptable to modern-day passengers were achieved.*

Apart from these hiccups, the Central was an undoubted success and even its foul air did not deter passengers from piling onto the system. Unlike the rather gloomy City & South London, the Central caught the public imagination, as shown by its confident publicity material featuring affluent looking passengers. The *Daily Mail*, owned by Alfred Harmsworth – an enthusiast, like the Prince of Wales, for newfangled gadgets – helped the process of generating public support by providing considerable positive coverage.

After much initial debate, the company followed the example of the City & South London by charging a single fare of 2d, irrespective of the length of the journey, and the original idea

---

* Crossrail, in contrast, will benefit from an incredibly sophisticated system of ventilation, which is designed not just to keep the temperature steady but also to divert smoke away from passengers in the event of a fire.

of having two classes of carriage was dropped. The railway became known at the 'Tuppenny Tube' and was an instant hit with the public, helped by the fact that it was pleasingly cool in the tunnels during the hot summer of 1900, the last of Queen Victoria's long reign, when temperatures regularly hit the 90°F (32°C) mark. In contrast, the Central boasted that below ground the temperature rarely exceeded 55°F (13°C). Indeed, in the winter, the relative warmth of the tunnels, where the temperature was pretty much the same as it was in the summer, also attracted people to venture onto the Tuppenny Tube.

The number of travellers using the Central Railway in its early days was impressive, soon reaching around 125,000 daily, and in fact overcrowding rapidly became a major problem. It was particularly acute at the Bank end in the mornings which prompted one wag, with rather more boldness than common sense, to suggest that both lines should be used to bring people eastwards in the morning and westwards in the evening, rather ignoring the problem of how to get the rolling stock back to run more trains since the trains would all pile up at a terminus.

Amazingly, thanks to the healthy number of passengers, the disaster with the rolling stock and the concerns over ventilation did not result in the company going bankrupt. Quite the opposite. The company prospered, paying its shareholders generous 4 per cent dividends in each of the first five years, not bad for such a risky venture involving a massive capital outlay. The private sector would never be able to do that today.

The Central had struck lucky. It had built a high-standard line on an existing busy transport route at a time when the alternative technologies, notably cars and motor buses, were not well developed and the main potential rival, the tram, had been banned. Its success lured investors to support other similar initiatives. The success of the Central Railway made it easier for the promoters of other Tube lines to obtain finance and remarkably, within less than a decade of the opening of the Central Railway, London had three new Tube lines. These were all built by an extraordinary – and undoubtedly crooked – entrepreneur, Charles Yerkes, an American who fetched up in Britain in mysterious circumstances around the turn of the century and managed to acquire control over a wide range of Underground interests.

Yerkes first gained control of the District Railway (then officially known as the Metropolitan District Railway) and electrified it, which greatly improved its efficiency since until then it had been operated by steam engines. Then, in an amazing financial coup, in 1900 he acquired three nascent Tube lines which had obtained parliamentary approval to be built but were struggling to find the funds: the Charing Cross branch of the Northern Line, at the time known as Charing Cross, Euston and Hampstead Railway; the Baker Street & Waterloo, which, following a suggestion in the *London Evening News*, assumed its current name of Bakerloo; and the Piccadilly, which had gone under various incarnations, as evidenced by the fact that to this day some of the stations still bear the letters G.N.P. & B. Ry – i.e. Great Northern, Piccadilly and Brompton Railway. Remarkably, all three managed to open their initial sections between March 1906 and June 1907, undoubtedly the golden period for underground railway building in London. It is not clear precisely where Yerkes got the money to fund this amazing burst of Tube line construction and he did not even live to see the completion of these lines, dying of a kidney disease at the end of 1905. The cost of building these lines was estimated to be about £370,000 per mile – about £42m in today's money.

The feat was all the more remarkable in that central London would acquire no more Tube lines for a further sixty-one years, when the first section of the Victoria opened in 1968, followed by the Jubilee Line in two sections starting in 1979 and completed on the last day of 1994. This was partly because Yerkes's heroic efforts meant that London had a surfeit of underground lines by the time all his projects were completed, but also because rival, road-based transport technologies were developing fast. Furthermore, the rigours of the First World War – and the economic austerity that followed it – ensured that no new Tube line in central London could possibly attract the private-sector investment which had provided all the funding up to that point. Britain's highly centralized political system meant that central, rather than local, government would be the only realistic source of funding for new lines and, for various reasons, this was not forthcoming during this protracted hiatus.

The various older lines did, however, expand further into the suburbs during this period. The Central, for example, reached West Ruislip in the west and Epping in the east by 1949 as a result of the New Works Programme of 1935–40. The eastern extension and adoption of a branch north of Ilford from the Great Eastern Railway had been partially completed before the Second World War (1939–45) but only opened to Ongar and Epping after the war. The Northern and the Piccadilly both expanded considerably, although some schemes, such as the idea of extending the Northern through Mill Hill East and on to Bushey, never materialized after being delayed by the Second World War.

None of these lines was built to the same scale as Crossrail as all were designed for the much smaller Tube rolling stock. There was one odd exception, and that was the Great Northern & City, a line built in the mid-1890s between Finsbury Park and Moorgate. The original intention had been to continue it on to south London, but sufficient funding was never forthcoming, in part because the diameter of the tunnel was 16 ft (4.9 metres), which meant a mile of tunnel cost twice that for the Tube lines. Little used, its ownership was passed around until it ended up with the Metropolitan Railway and it is now, thanks to its size, part of the national rail network rather than belonging to the London Underground. The Great Northern & City was, however, a missed opportunity to create an early deep underground route that traversed London and would have been accessible to mainline trains.

The increasingly overburdened small-bore Tube lines have been the mainstay of London's Underground transport system. However, anxieties about the prohibitive expense of widening their tiny tunnels while maintaining a service on the line, and the absence of an overall plan set out by government and backed by appropriate subsidies, meant that the measures needed to relieve pressure on the east–west Tube network were a long time in coming. Crossrail was also envisaged initially as a conventional Tube, but, happily – in the face of the chronic overcrowding on the existing network – common sense would prevail.

2.

The Crossrail concept

The need for better transport links between west and east London had been recognized as far back as the opening decades of the nineteenth century. The first major transport connection between the two was the Regent's Canal, which opened fully in 1820, enabling goods to be transported east from the Grand Junction Canal at Paddington through to the Thames at the Limehouse Basin. The canal was punctuated by several major basins where goods could be loaded and discharged, notably at Battle Bridge just north of the site of King's Cross station, and City Road, which served the City of London. A late addition to Britain's waterway network, the Regent's Canal enjoyed a short-lived heyday, the tonnage of coal it carried being boosted briefly by the opening of the London & Birmingham railway in 1838. However, it soon became clear that the railways – with their much greater speed and consequently reduced costs – were the future. Sensing that the canal was no longer financially viable, its owners began trying to promote the idea of converting the Regent's Canal into a railway. In 1845, they attempted to sell the waterway to a group of businessmen who had formed the Railway Canal Company for this purpose. Their prospectus stated that 'by the proposed railway, passengers and goods will be brought into the heart of the City at a great saving of time and expense, and facilities will be afforded for the more expeditious transmission of the mails to most parts of the kingdom'.[1] Like so many Victorian projects, it was not to be. And two further attempts to convert the canal into a railway over the next couple of decades also failed, in part because the government objected to the idea of a railway passing through Regent's Park.

The existence of the North London line, circumventing central London along a semicircular route through its northern suburbs, and, from 1863, of the Metropolitan Railway – both of which carried goods trains – made it difficult to attract capital for those seeking to convert the canal into a railway. While there was undeniably a market for rail freight in east London to serve the docks, there was less demand for passenger trains because most of the thousands of stevedores lived locally in the East End and had no need of the railway. They were well served by local trams, which were much cheaper than trains. As a consequence, the Underground system did not stretch into east London and its docks until the expansion of the District Line in the first few years of the twentieth century.

Nevertheless, efforts to convert the Regent's Canal into a railway continued when, in 1883, the canal was sold to investors whose intent was revealed by the new name they gave to the company: the North Metropolitan Railway and Canal Company. Despite this, the idea of conversion was soon dropped as being impractical and, instead, the canal continued to be used to transport goods, although carryings never regained the level achieved in its heyday.

London's east–west rail connections therefore remained inadequate, and the service for passengers deteriorated as London's population expanded dramatically on its east–west axis in the second half of the nineteenth century with no concomitant increase in rail capacity. It was a gap in provision long recognized by transport planners but a difficult one to solve.

After Yerkes's successful construction of three new Tube lines, central London was very well served by the Underground. Consequently, the focus of the Combine, the huge transport grouping which encompassed all the Underground lines, except the Metropolitan, and many bus services, and became the core of London Transport when it was formed in 1933, was to extend existing lines into the suburbs. Edgware in the northwest, for example, was reached in 1923, Morden in south London in 1926, Cockfosters in 1933 and Uxbridge in 1938. There was, though, little interest in connecting the two sides of London and, in fact, one of the few north–south connections, the Snow Hill Tunnel between Farringdon and Holborn Viaduct, opened in 1866, was closed for passenger traffic during the First World War. The prevailing view of planners in this period – that people would travel for work *into* central London but would have little inclination to travel *through* it – perpetuated the radial nature of the network serving the capital.

The issue of improving east–west links was not taken up again until the early 1940s, when plans for the future of London after the conflict were being mooted. A new east–west cross-London Underground line was mentioned in both the County of London Plan, published in 1943, and the Greater London

Plan, known as the Abercrombie report, issued the following year, but these were only vague references buried among a lot of hoped-for transport projects which concentrated mainly on improving the roads. The County of London plan did, however, include a map of proposed rail projects which shows a line running with a rather odd curve southwards to Cannon Street and then through to Victoria with a spur going to Paddington.

In 1949, a remarkably ambitious scheme for transforming London's rail network was published by the British Transport Commission, the government agency which controlled all public transport, British Railways and British Road Services. The London Railways Plan envisaged a series of cross-town lines running under London in tunnels designed for full-size railway carriages which would also have ensured they could be used for freight. A total of thirty-four miles of tunnel would have been built under the capital, linking, for example, Fenchurch Street services with Waterloo and the Cambridge lines into King's Cross with Victoria. The plan did not include an equivalent of today's Crossrail, though it did envisage a route linking Euston with Blackfriars. Of course, there was no serious consideration of such a massive scheme, let alone any funds for it, and consequently nothing came of the plan. At a time of post-war austerity, the proposed cross-town railway plans remained no more than lines on a map.

This was a period during which transport planners were entirely focused on catering for motor cars and lorries. The idea of building motorways was already being discussed, and the first section of what is now the M6 (the Preston bypass) was completed in 1958 and parts of the M2 in Kent soon followed. The railways, along with the Underground system, were seen as a redundant and dirty technology, irrelevant to the modern post-war world. As a result, both entered a long period of decline after the war in the face of competition from road vehicles; in the 1950s there was more talk of shutting Underground lines than opening new ones. When the Victoria Line was given the go-ahead in the early 1960s, it was only as a result of fears of growing unemployment, rather than because of any perceived strong need to improve London's rail and underground network.

It is not surprising, therefore, that the very broad plans for a Crossrail-type line were left in abeyance for such a long time. Crossrail's gestation period of seventy years between its initial conception in the 1940s and its eventual go-ahead can be explained – if not justified – by the changes in transport planning over that period rather than the oft-mentioned failings in the overall planning system for big projects. London in the 1960s was set to become the UK's principal testing ground for policies based on improving road conditions in urban areas. Already, in the early 1960s, Park Lane, a three-quarter-mile-long north–south thoroughfare on the western side of London's Mayfair, had been turned into a dual carriageway at the massive cost of £1.15m (about £25m today) and the loss of 20 acres of Hyde Park in a scheme justified by a transport minister of the day, John Boyd-Carpenter, because 'no single road development scheme could make a greater contribution to the relief of growing traffic congestion'.[2] Much worse was envisaged. A plan first mooted in the wartime reports on London's future was to create a series of ring roads in and around London to ease traffic flow. These proposed roads became known as the motorway boxes and the innermost one would have destroyed large swathes of central London as it would have run through a series of inner suburbs such as Shepherd's Bush, Hampstead, Bethnal Green, Blackheath and Battersea. Construction was started on part of the western section at Shepherd's Bush,* and a spur into central London to Marylebone, now known as Westway, was completed in 1970 (and opened by the transport minister Michael Heseltine). Fortunately, however, the overall plan was scrapped by the Labour-controlled Greater London Council (GLC) in 1973, which bowed to the growing weight of opposition. Consequently, London escaped the fate of other British cities, notably Birmingham and Glasgow, as well as smaller towns such as Gloucester and Hereford, where the motor car was allowed to rule supreme and where swathes of housing and industrial buildings were sacrificed to make way for roads which cut through entire districts, fragmenting and destroying communities.

There had to be other ways of solving the increasing pressure on the centre of London and the election of a Labour government in February 1974 saw the start of a search for alternatives. However, it would be an over-simplification to suggest that Labour inevitably favoured public transport solutions while the Conservatives pushed for more roads. Although this was the general pattern in historical terms, there have been some significant exceptions: the Conservatives spent the staggering sum of £1.24bn (more than £30bn today) in the mid-1950s on a Modernisation Plan for the railways, while in the 1960s the new Labour government of Harold Wilson continued the programme of rail closures post-Beeching (despite having made promises to the contrary in their election campaign). Similarly, while party

---

* A redundant little spur less than a mile long, it was named the M41 and is now the A3220.

politics inevitably plays a role in the Crossrail story, it is not always obvious who is on which side.

At least, though, rail was back on the agenda as a potential solution to the problem of overcrowding on the roads. A report on the future needs of London's railways, the *London Rail Study*, the first of several such works, was produced in 1974 by a joint team from the Department of Transport, British Railways and the Greater London Council and suggested an ambitious scheme of building two rail lines under London. The study was the first to mention the term 'Cross-rail' and although it recognized that numbers travelling by rail in London were actually falling, it used innovative computer-based studies for the first time in the UK to make the case for new railways. The report pointed out that, despite the overall fall in passenger numbers, parts of the network were extremely crowded at peak times and needed new lines to relieve them. Moreover, it argued confidently – and correctly – that the decline of the early 1970s would be reversed, though its authors certainly failed to anticipate just how rapidly rail travel in the capital was going to grow over the next four decades.

The study recommended 'two BR Cross-rail lines', which were to be interlinked. The first was to run between Paddington and Liverpool Street, connecting at both Holborn and Leicester Square with the second line, which would been between Victoria and London Bridge. The report also set out plans for what is now Thameslink (which would be achieved a few years later by the simple expedient of reopening Snow Hill Tunnel) and for the 'Fleet Line', which would later be built as the Jubilee Line (though not, as anticipated, serving Fleet Street). The report also endorsed the idea of a line between Chelsea and Hackney, a variant of which is now being considered as Crossrail 2.

One of the reasons why the Crossrail schemes were deemed necessary was that London's natural growth pattern was constrained by the Green Belt policy, a concept devised as far back as 1935 but principally brought into force in the aftermath of the Second World War. This recognized that if London employment levels were to continue growing, commuters would have to undertake longer journeys. Crossing London, however, remained difficult, involving a change onto the Underground system and, quite possibly, the need to take another train on the other side. The Crossrail concept was a way of relieving this, offering the prospect of long journeys with possibly just one change or none at all.

The 1974 study estimated that 14,000 passengers would be carried during the morning peak between Paddington and Marble Arch and 21,000 between Liverpool Street and Ludgate Circus. The study also suggested there should be through services to Heathrow Airport, which, at the time, had no rail connection and was only linked with the Underground system three years later, in December 1977. The estimated cost was £300m – even accepting subsequent levels of inflation, this was remarkably optimistic. Although a feasibility study for the scheme was recommended by its authors as a high priority, nothing was done.

The mid-1970s were not a good time for visionary new schemes to add to London's already extensive rail network. The recommendations of the 1974 *London Rail Study* were largely ignored as the consequences of the financial crisis, and particularly the high inflation rate, triggered by the oil price rise of 1973, created an atmosphere that was hardly conducive to building major projects. London's population was also shrinking and use of the Underground had plummeted. While railway managers were putting up ideas for new lines, politicians were sceptical of the value of the rail network and were making proposals to close lines and cut back on services. But at least the first part of the Fleet Line – as the Jubilee Line was initially known – had been under construction since 1971. This was a relatively cheap scheme since it only involved a 2.5-mile (4-km) stretch of new line from Baker Street to Charing Cross via Bond Street and Green Park, and the separation of the two legs of the Bakerloo, with the Stanmore branch becoming part of the new Fleet Line. At the time it was expected that the second section, running under Fleet Street via Aldwych and through to Cannon Street and Fenchurch Street, with a possible extension to Lewisham, would quickly follow. In the event, the first section of the renamed Jubilee Line was not completed until 1979, and the second, on a different alignment which bypassed Charing Cross, leading to the abandonment of a few hundred metres of tunnel including a never-used new section that had almost reached Aldwych, not until the last day of 1999.

The 1974 study scored a massive own goal by playing down the need for major new schemes: none of the ideas for extending London's rail network could, it asserted, 'be justified either on financial grounds or on a conventional social cost/benefit assessment of their transport effects'. Indeed, it is salutary to note that various sections of railway line in London were closed in the wake of the Beeching report of 1963. The North London line, which can be considered a cross-London line linking east and west, was actually slated for closure by Beeching and only saved after a lengthy campaign. Even then, it was run in a half-hearted way – first by British Rail and later under privatization in 1997. However, it would later become well patronized following considerable investment by Transport for London (TfL) in its track, stations and trains.

In view of the pessimism of the 1974 study, it was not surprising that, apart from continuing work on the Fleet Line, it took another decade and a half for renewed interest in major railway projects for London to re-emerge. This time, serious consideration would be given to the concept of Crossrail. Overcrowding and massive passenger dissatisfaction with both the Underground and London's mainline stations had begun to create the need for the discussion of large-scale railway investment projects in the capital. Use of the Underground system had reached a post-war low in the early 1970s, with fewer than half a billion passengers (compared with the 2016/17 total of 1.37bn), but then began to recover. The policy of pushing offices out of London through the Location of Offices Bureau* – which, ironically, was advertised in Underground carriages to promote the concept – was abandoned and employment numbers in central London began to creep up after a long decline.

In November 1980, British Rail, then served by one of its best chairmen, Peter Parker, produced another forward-looking discussion paper entitled *A Cross-London Rail Link*, setting out plans for a variety of railways under or through London. As Parker, showing a penchant for alliteration, put it in the introduction, the hope was to 'make a journey from Peterborough to Portsmouth, Watford to Woking as straightforward as the London Underground already makes it from Barnet to Balham'.[3]

---

* The Location of Offices Bureau was a quango set up in 1963 to disperse office jobs from central London and abolished by Margaret Thatcher in the 1980s.

There was no shortage of ambition in Parker's paper: his plan was to send mainline express trains through the capital as it would 'enable an extension of the quality of Inter-City travel, already established on the principal routes to the north and west, to the main lines south of London'. Various alternative routes were put forward, including Clapham Junction to Paddington via a low-level Victoria station and Waterloo to Euston, all requiring, as the authors recognized, the construction of deep-bore tunnels under the capital. One suggestion, put forward in an innovative pamphlet, was that the Birmingham services could be linked with the Brighton trains to provide a through route, enabling passengers to avoid using the crowded mainline stations, and reducing the journey times between provincial destinations – just as the partially built M25 would eventually achieve for motorists. The trains, it was recognized, would have to be electric and consequently dual-voltage since there are different systems south and north of the Thames. Optimistically, the pamphlet said that some 200 fewer trains would be needed as they would no longer have to spend dwell time at the London termini. However, the pamphlet failed to address one fundamental problem for such cross-London railways, namely that the requirements of Inter-City-type trains (seating suitable for long-distance travel, impressive top speed, restaurant cars and toilets) are very different from those of trains used for short suburban journeys (fewer seats, a greater number of doors for easy entrance and exit, no facilities, engines built for lower speeds and frequent stops and starts).*

---

* The trains selected for Crossrail will follow this pattern and will therefore have no toilets, saving a huge amount on maintenance and servicing, but undoubtedly inconveniencing some passengers..

Nothing came of either of these innovative reports. The early 1980s were still not a propitious time for British Rail, given the hostility of Margaret Thatcher's new Tory government towards the railways. However, there was a renewed interest in railway development in London later in the decade as commuter numbers rose. This was the stimulus for the preparation of another *London Rail Study*, a report produced by a group of civil servants from the Department of Environment and the GLC to examine future rail-based transport needs in the capital and the South East. Richard Meads, a planner with London Transport, says that soon after the King's Cross fire of November 1987, 'we began to recognize that there was a need for new capacity on the Underground and we were asked to produce suggestions'. One of these was an RER-type railway for London and the idea of an east–west tunnel under London was mooted. Meads says he is not quite the 'father' of Crossrail, 'but I was certainly one of those pushing for it at a very early stage as we produced a report in March 1988 which included a version of Crossrail'.[4] This led to the establishment of a more formal committee to produce the *Central London Rail Study*, which put forward several suggestions for new lines and extensions, including an east–west route that it dubbed 'Crossrail'. The thinking behind the study was influenced by the development of rail services in Germany and France that went across town in a tunnel, linking the suburbs on either side. In France, the first two lines of the RER were completed in 1977, bringing to fruition a concept that had first been set out in 1936 in a plan for a 'metropolitan express' connecting suburbs with a railway under central Paris. This was delayed by the war and the need for subsequent reconstruction, but already in the 1960s the French authorities envisaged a series of lines crossing Paris in tunnels and spreading out far into the capital's suburbs. After a construction period of eight years, two lines, A and B, opened on the same day (9 December 1977) linked at Châtelet–Les Halles.

At 2018 prices, the construction of the RER cost around €2bn – spending on a far grander scale, in proportion to the local population numbers, than had ever been undertaken in London. It was made possible by the *versement transport* (VT), a tax introduced in 1971 on local employment which was initially conceived as a means of raising investment for transport infrastructure projects, but is now increasingly used to cover the operating costs of the RER. Thanks to such a generous source of revenue, the new stations for the RER system were built to a far grander scale than was strictly necessary. This was a cause of some controversy, but was justified by the architectural merit of the stations and by the need to avoid the claustrophobic feel of many parts of the Métro. The new RER stations were akin to underground cathedrals, built as single halls housing the two platforms and three times the length of traditional Métro stations. The tunnels, too, were built to larger dimensions than those of the Métro – a prescient course of action as double-decker trains are now used extensively on the network. The scale of the RER was in marked contrast to that of London's Victoria Line, built almost simultaneously but which now suffers from chronic overcrowding because of the lack of foresight of the planners. Stations on the Victoria Line frequently have to be shut at peak times because of overcrowding on the platforms, despite the fact that, thanks to new trains and modern technology, the line now has a frequency of up to thirty-four trains per hour. It will be impossible to add more capacity to the Victoria Line, since its stations could only be expanded at enormous expense. This experience has informed the Crossrail planners and explains many of their decisions.

The early RER lines, too, proved to be far more successful than anticipated. Despite the generous space standards to which they were built, RER stations are overcrowded at peak times, a situation made worse by their use of double-decker trains with only two, relatively small, entrances on each carriage. Within a few years, Ligne A in particular was full to capacity at peak times, with a flow in the central section of 55,000 passengers per hour in each direction, one of the heaviest in the world. The success of the RER was partly a result of the way that the new lines altered the habits of Parisians, demonstrating that rail lines can be a catalyst for change. The existence of the RER made the *banlieusards* – as suburban dwellers are called – much more likely to travel into the centre of Paris, especially for leisure and shopping purposes, and helped the process of revitalizing the inner city that has now become a common phenomenon in many first world cities. The newly connected suburbs, too, flourished as a result of the new railway service. By the end of the twentieth century, three further lines had been built, and several extensions are now being envisaged.

In Germany, a similar system of railways, known as the S-Bahn (*Stadtschnellbahn*, or city express railway), had its origins in the interwar period in Berlin with lines that originated in the suburbs and crossed the city. However, in many respects these were indistinguishable from the underground lines in London as they were built on the same scale as the U-Bahn (underground) lines. It was actually the remarkable network of lines built in Munich that caught the attention of London's railway planners.

Plans for an S-Bahn in the city had been drawn up under the National Socialists and a start had been made in 1938. However, the outbreak of war delayed the implementation of the plan until 1965 when a decision to build the scheme was finally made jointly by the federal, regional and local authorities. The need for a more efficient railway network was given further impetus the following year when Munich was selected to host the 1972 Olympic Games. The system was completed in time, but the network presented a level of operational challenge that left its mark on the visiting London planners. As with Crossrail, there is a central tunnel but there are seven branches on the western side and five in the east. This means that trains from a very wide variety of destinations, coming from lines which may be suffering all kinds of operational issues and delays, have to feed in through a single two-track tunnel in a pre-set order. In Munich, the frequency of trains on the outer branches is only three per hour but this still generates a level of operational complexity which would not, as we will see, be possible in the far more heavily used suburban services in London. Hamburg, which was also mentioned specifically in the study, was then in the process of constructing a tunnel linking the two sides of the city.

The UK was clearly lagging behind in developing the concept of these cross-town railways. Actually, there had been a suggestion by British Rail to create precisely such a network in Manchester in the early 1970s. Don Heath, an engineer who was later employed by Crossrail, worked in Manchester at the time on a plan to link Victoria and Piccadilly stations with an underground tunnel which would then have carried trains out to the suburbs: 'Unfortunately the plan collapsed,' he says, 'first because British Rail delayed it by asking for more detailed estimates and then Newcastle came in with its plan for a metro, and that meant the money was not available.'[5] The UK, therefore, would not get its first underground cross-town railway for another forty-five years.

It was overcrowding that prompted the next report on London's railways, which would lead to the first attempt to build Crossrail. The backdrop to the *Central London Rail Study* of the late 1980s was soaring employment levels in central London. After decades of decline, when the focus was on the suburbs, inner London was being revitalized with large developments stimulated by what are now termed as agglomeration effects – the fact that people in particular industries prefer working alongside each other, even when the companies concerned may be rivals. The big stimulus for central London employment was Big Bang, the relaxation in 1986 of rules for conducting business on the Stock Exchange, which led to a large increase in employment and the consolidation of many big players, all of whom wanted to be in the Square Mile.

The London Underground was creaking under the strain of increased demand and was entering a period of recovery after the King's Cross fire of 18 November 1987. That disaster, in which thirty-one people were killed, was the result of a fundamental malaise in the operation of the London Underground. A fire that had been smouldering under an escalator for half an hour suddenly erupted into a fireball, costing the lives of the people on it. The subsequent report by Desmond Fennell QC exposed a series of shocking practices in the operation of London Underground symptomatic of an organization in decline. The most likely cause of the fire was thought to have been a smoker's match which had fallen through a gap beside the stair tread into the bowels of the escalator where junk, much of it inflammable, had been allowed to accumulate for years. Not only did the smouldering fire remain unnoticed, but the number of staff on duty, who might have detected it, was depleted because of a culture in which routine 'bunking off' work early or being signed in by colleagues despite not reporting for duty had become the norm. Fires, or 'smoulderings' as they were called in an obvious attempt to disguise their significance, were considered routine and unavoidable. Training in emergency procedures was non-existent and the management was sloppy and remote. In other words, the fire had been eminently preventable.

More importantly, the King's Cross disaster was a wake-up call to the transport planners and the politicians charged with making investment decisions. London's transport facilities were in dire need of an upgrade and there was now widespread recognition that they had to be improved and modernized. Another long-term effect of the fire was the raising of the standards expected in building or operating railways underground. A huge raft of measures were now considered to be necessary to ensure the safe operation of the Underground, and this imposed costs on both the running of the system and on the construction of any new lines. When glib comparisons are made between the costs of constructing the early Tube lines with those of a modern railway like Crossrail, the additional safety measures, now mandatory, have to be borne in mind. As we shall see, everything requires back-up and every risk has to be assessed and reassessed. Indeed, today, the safety culture is built into every aspect of a project like Crossrail and that imposes huge extra costs. The events leading up to the King's Cross fire would simply not occur today and, even in the unlikely event that a fire began under an escalator as it did during that fateful evening in 1987, it would be rapidly detected and extinguished before any major damage were caused.

The need for more investment on both the Underground and other London railways, together with the realization that Big Bang was stimulating a rapid growth in employment in the City, led to the commissioning of the *Central London Rail Study* in 1988. Its launch was the result of an initiative by the Department of Transport, but it gained greater resonance from being a collaborative effort involving British Rail as well as the London Underground. Its remit was to 'develop a strategy for improving services for passengers on the British Rail and Underground networks in London, and to provide forecast demand on the two networks to the end of the century, with particular reference to passenger congestion in the area bounded by the major rail termini and their approaches'. Overcrowding on the existing railway allied with the government's belief that, after a period of decline, London would now grow fast again, were the key drivers.

The study, which was produced remarkably quickly and published in January 1989, set out the figures for recent trends in passenger numbers. It argued strongly that without considerable investment London's transport system would soon be unable to cope and would become an impediment to economic growth. On the Underground, daily passenger numbers in the morning peak had increased from 305,000 in 1980 to 415,000 in 1988, representing a rise of 36 per cent. As the study stated, 'both London Underground and British Rail are carrying record numbers of passengers into central London'.[6] The study noted two trends that pointed to a Crossrail-type solution. First, it showed that half the British Rail passengers arrived between 8 and 9 a.m., putting particular strain on the system at those times. Overall, the Underground was having to cope with a 50 per cent increase in the number of rail passengers transferring onto the system from their commuter trains and the only way to reduce these numbers would be through direct services from the suburbs into central London, obviating the need to transfer at a mainline station. Secondly, because the Green Belt effectively increased the distance many of the growing number of people employed in central London had to commute, the study found that 'the tendency for growth to be focused on longer distance commuting from beyond the central London boundary has also led to acute problems on certain services'.[7]

It was not just the mainline stations that were under pressure. The report stated, 'at some 25 Underground stations, serious congestion is already occurring at peak periods, either in the ticket hall, on lifts or escalators, or at platform level. A further 15 stations will become seriously crowded by the end of the century if demand continues to grow as forecast.'[8]

The study assessed an impressive list of potential projects, many of which had featured in the 1974 study, including the Tube line linking Chelsea and Hackney (two areas not served by the Underground), and an extension of the Jubilee Line out to Ilford via Whitechapel and Stratford – pretty much the ultimate alignment of Crossrail. These were envisaged as Tube lines and, rather more excitingly, in line with the RER and S-Bahn thinking, the study set out three potential 'Crossrail' routes. The first would have linked local services running to and from Euston and King's Cross with southern services to and from Victoria, relieving all those three stations; the second was an east–south

Crossrail which would have linked the services to and from Liverpool Street with those serving Victoria. These two would probably have been mutually exclusive given the pressure on Victoria. In the event the route most favoured by the 1989 study was an east–west line, but it was somewhat different in concept from the route that later emerged. Like the present scheme, it would have linked the Great Eastern suburban lines serving Liverpool Street with those out of Paddington, but there would have been a junction somewhere under Hyde Park with a line running up to Baker Street and Marylebone, to take in services on the Metropolitan Line and Chiltern services. The idea for this junction came from the recognition that the potential demand for these cross-London links was lopsided: whereas there was an intense and heavily used suburban service running into Liverpool Street, Paddington, in contrast, has relatively few suburban services. This is the result of history. As mentioned in the previous chapter, the Great Western Railway built in the 1830s had few stops serving the villages that would later become the western suburbs of outer London; whereas on the eastern side, much of which was constructed rather later, numerous stations were built and served by cheap workmen's trains. This imbalance between east and west, as we shall see, remains a problem today.

Jim Steer, a transport consultant who has been involved in several aspects of the Crossrail story, believes that the mistake made by the authors of the study was that they concentrated too much on central London (defined as Zone 1) rather than inner London as a whole: 'The planners' view at this stage was that central London was booming, and consequently they ignored everything else. So they suggested tunnels under London but then would do little to provide for the outer areas, going out of the tunnels at the cheapest point rather than going where they might serve local residents better.'[9]

In the opinion of Michael Schabas, another veteran of numerous London transport projects, the study failed to take a long-term view of the likely changes in London: The *Central London Rail Study* had no larger "strategy" ... There was no attempt to open new development sites, either for jobs or homes. The study tested schemes against a "fixed trip matrix". In other words, it assumed that the jobs and homes would be in the same locations, and the same origins and destinations [of journeys] would be linked, whether or not the proposed investments were built.'[10] There was, in particular, little consideration of Docklands which would, as we shall see in the next chapter, play a major role in this story.

3.

Megaprojects and mega-businesses

With the palpable overcrowding of London's transport network and the probability that the capital's economy would continue to grow rapidly, there was an expectation among transport planners and local politicians that the best of the schemes in the Central London Study would quickly be waved through. However, even though the need for a solution was urgent, the size of the proposed schemes meant that bringing any of them to fruition would be a tough ask. Rational and sober argument would not be the sole factor driving the process of decision-making; politics, too, would inevitably play their part. And thus it was that the first effort at planning Crossrail would end in tears.

The assessment of the value of major transport projects is carried out through a methodology called cost–benefit analysis, which is supposed to determine priorities. The idea is that the benefits and costs of a scheme can be assessed and then set against each other to provide a ratio of the two. Then the funder, normally the government from one pocket or another, can judge whether it is worth going ahead. A scheme with a ratio of one, therefore, is pretty poor since it means that the benefits are equal to the costs, suggesting that the scheme might not be worthwhile. Anything below one is therefore a no-no and unlikely to attract funding, while scores of two or above are considered an excellent basis for a project. The benefit–cost ratio then becomes a key component of the 'business case' on which schemes are judged. Having a good business case essentially means having a healthy benefit–cost ratio.

The rules for the assessment of projects are set out by the Department for Transport in a fiendishly complex system known by the unfathomable name of WebTAG (Web Transport Analysis Guidance), which is updated occasionally to reflect new thinking. In truth, cost–benefit analysis is really art masquerading as science, since the assumptions that underpin the analysis are pretty random and can be manipulated to get the answer that the client, or the consultants who invariably carry out these studies, sought in the first place.

There are problems on both sides of the equation. In relation to costs, there is always a high degree of uncertainty, which have famously and accurately been described as 'unknown unknowns'. The bigger and the more complex the project, the greater the likelihood that something will go wrong and that costs will soar. In transport projects, unexpected ground conditions are the most common cause of major increases but, in fact, they are so frequently found that they can hardly be described as 'unexpected'. In addition, there are many other uncertainties, ranging from the vagaries of the demand for construction workers to the availability of the best contractors. The Jubilee Line Extension suffered particularly badly from a late increase in costs as some of the workforce realized that with the deadline set by Prime Minister Tony Blair of the last day of 1999, they could extract large bonuses from London Transport. In the words of Michael Schabas: 'The trade unions took advantage of this, with the electricians, installing critical equipment, holding a ten-day strike in late 1998.'[1] Setting a firm deadline always creates a hostage to fortune and, as we shall see, while Crossrail has committed itself to a staged series of openings, the project was at one point extended for a year in order, interestingly, to save money.

On the benefits side of the equation, the key problem is one of definition. While the fares that a scheme generates of course make up a substantial proportion of the total, they are rarely sufficient to justify a scheme going ahead. Indeed, if they were, the private sector would evidently be quite happy to finance a scheme entirely, knowing shareholders would receive a healthy rate of return. Major schemes, however, are rarely, if ever, in that happy situation, owing to the scale of the funds required for investment, the huge risks of cost overruns on big projects and the length of time for any profits to come through. Therefore, in order to justify a scheme, other benefits, which are not directly or wholly financial, have to be put into the analysis. In transport schemes, the principal 'benefit' other than fares income normally consists of the time savings made as a result of the new scheme. If a passenger can get to their place of work twenty minutes quicker, then that is assumed to make them more efficient. These time savings are not confined to the passengers using an extended rail service or the motorists on a new road, but can also include those made by non-users through the overall reduction in congestion that may result from a major road or rail scheme. A more recent addition to the methodology has been what are termed 'agglomeration effects', the advantages gained from having several companies in the same industry sited close together. This has been demonstrated over time to lead to considerable savings as a result of greater specialization, exchange of knowledge and development of a skilled labour force. A transport scheme, therefore, may well have the effect of making it possible for businesses to move closer together and this is reflected in the cost–benefit analysis. Ultimately, so says a civil servant who cannot reveal his name, 'the inclusion of the benefits of these agglomeration effects swung it for Crossrail and when they were set out in a document, the Government, the Treasury and Transport for London all became supportive of the scheme'.[2]*

All these benefits and savings can – with great difficulty and with some dubious assumptions – be ascribed a monetary value, with different amounts for, say, gains in leisure time compared with those made by people on business, or for a senior manager as compared with a clerk. This methodology, which has spawned a very extensive and lucrative industry for a small number of consulting firms, is necessarily haphazard and often given far more prominence and set out in far greater detail than the necessarily arbitrary calculations can conceivably justify. It is a useful way of comparing the potential of different projects rather than their intrinsic value. As such it is a perfectly valid methodology allowing the most favourable scheme between various alternatives to be chosen. However, this methodology has now become absolutely standard with the value of the benefit–cost ratio accorded far

---

* See also Chapter 7, p. 153.

too much significance. In particular, press reports that suggest a particular scheme will be worth '£xbn' by 2050, are highly misleading, presenting speculative and in some cases spurious figures as though they were established facts. Anyone looking back from 2050 and assessing the benefits will almost certainly discover the numbers were completely wrong.

In recognition of the vagaries of this process, the Treasury now insists on imposing an extra 30 per cent on the costs side, a calculation known as 'optimism bias'. That is to reflect the fact that in the past many megaprojects have gone over budget, completely negating the original benefit–cost ratio calculation. This presents an extra hurdle for schemes to be given the go-ahead. In the 1980s, however, the concept of 'optimum bias' had not yet been developed and the authors of the *Central London Rail Study* had the luxury of considering and comparing a wide range of projects. As ever, the way they came up with their recommendations was remarkably haphazard and was influenced by the politics of the time. Ultimately, the final decision was down to the politicians, and the zeitgeist dictated that market forces should determine which developments were approved, even though the final bill would largely be paid by the public through their taxes and by passengers through the farebox.

The *Central London Rail Study* found that the best option, in terms of the benefit–cost ratio, was the east–west Crossrail with a tunnel large enough to take mainline trains, rather than only being able to accommodate the much smaller Tube trains. Oddly, little consideration was given to pursuing what was the cheapest and easiest cross-London link – the extension of Thameslink services. Thanks to the intervention of the GLC, trains had just started operating through the reopened Snow Hill Tunnel linking, for example, Bedford with Brighton and St Albans with Sutton. However, Thameslink's potential was not really exploited: even at peak times there were only eight trains per hour in each direction, despite the fact that it had the capacity to serve a wide variety of additional destinations both north and south of the river. Although the Thameslink extension had 'benefits twice the costs', the authors of the study seemed to find the project insufficiently exciting: 'it is not a new line and its impact on congestion is limited.'[3]* The Chelsea–Hackney Line had the second-best business case, while the third-best, the Jubilee Line Extension, was disregarded because it trailed a long way behind as it was considered the least useful in relieving congestion.

Not surprisingly, even without 'optimum bias' being factored into the equation, the Treasury was deeply suspicious of these big projects – which are now dubbed 'transport megaprojects' – and subjected them to rigorous analysis. Its favoured tactic to avoid spending money was to announce further studies which would inevitably lead to delays, in the hope that they would all one day disappear, leaving the Treasury's coffers unraided. It took a strong politician – or politicians – to push a project through, and good luck with the timing of their efforts as well. The first time Crossrail ran out of luck because another project, the Jubilee Line, was more in tune with the zeitgeist but the second time it was more fortunate.

Research into megaprojects, those defined as costing billions rather than millions of pounds, provide some justification for the Treasury's scepticism by highlighting the widespread tendency

---

* It would eventually be adopted as Thameslink 2000 – an over-optimistic name since the scheme was not actually completed until 2019.

of promoters of new projects to understate costs and over-state benefits. In an analysis of figures produced by the World Bank, the Danish economic geographer Bent Flyvbjerg, in his book on megaprojects, found that 'cost overruns of 50 per cent to 100 per cent in real terms are common for large transport infrastructure projects, and overruns above 100 per cent are not uncommon'.[4]

On the other side of the equation, 'demand projects that are wrong by 20 per cent to 70 per cent compared with actual development are common'[5] for these schemes. In other words, not only do projects tend to be more expensive than expected, but they also attract fewer users than predicted. Flyvbjerg argues that there are structural reasons for this bias, which, since there are 'many more underperforming projects than can be explained by chance alone... cannot be explained primarily by the innate difficulty of predicting the future'.[6] Flyvbjerg blames the over-eagerness of politicians to make a mark in the short time they have in office and pressure from special interest groups and contractors for the over-optimistic view of megaprojects which in turn makes them more likely to be given the go-ahead. Once work on projects is put out to tender, there are considerable incentives to underestimating costs in order to win contracts, because the penalties for overspending are relatively small. Remarkably, the Major Projects Association, the industry's representative organization, has acknowledged that 'too many projects proceed that should not have done'.[7]

However, while some megaprojects have undoubtedly proved to be white elephants, with, in hindsight, little justification for their construction, many do produce considerable wider societal benefits when their overall impact is taken into account. More recent analysis, by the University College London Omega research team into transport megaprojects, is rather more supportive of the concept than Flyvbjerg, suggesting that the benefits of megaprojects need to be looked at in a wider context. While recognizing that overspending is a problem, the researchers found that there is a tendency to view projects such as Crossrail too narrowly as mere transport projects when, in fact, they are 'agents of change'. Many projects were evaluated 'without sufficient attention being paid to their potential capability to directly or indirectly stimulate urban regeneration and wider spatial and sectoral change'.[8]

Furthermore, such projects have to be allowed 'time to breathe', ensuring politicians do not make hasty decisions over their viability while they are still being constructed or shortly after completion. The Omega team stressed that transport megaprojects 'are "organic" phenomena rather than static engineering artefacts',[9] and thereby inevitably become 'agents of change'. Of course, some of this change, such as increased traffic from a road scheme or the destruction of an attractive environment, may be negative. Conversely, certain potential benefits may be more difficult to predict: unexpected outcomes, including new developments around stations or bigger than expected passenger flows on certain routes, may result.

The need for a wider assessment certainly applies to Crossrail. It is already possible to see positive outcomes from the project to which it is difficult to ascribe a value (such as its training of more than a thousand apprentices and the creation of a special college to train tunnellers at Ilford in East London*). The boost

---

* See Chapter 13, p. 277.

to London's self-confidence and international prestige are difficult to quantify, too. Technological innovation, on which the Crossrail team has focused particularly, is another potential – and unquantifiable – gain. Just as the US space programme resulted in numerous practical inventions, so Europe's largest infrastructure project offers the prospect of generating a wealth of new ideas. Crossrail's innovation programme, which its contractors were required to help fund, offered those involved in the project rewards for pioneering 'techniques, products and methods' that could benefit the entire construction industry in the future. These advantages cannot be quantified in the narrow bean counters' world of benefit–cost ratios.

One key disadvantage of the widespread use of the benefit–cost ratio methodology for transport projects is that road schemes often come out much better because their 'benefits' are so easily quantifiable. Consultants find it relatively easy to estimate future traffic numbers and multiply that by time savings for every vehicle occupant and every flow of goods in lorries. The benefits of rail schemes, whose advantages may be environmental or less tangible in other ways, are harder to evaluate. What's more, there is a certain confusion about the way new road schemes are assessed. They often induce people to use their cars instead of other means of transport or not travelling at all: to describe this as a 'benefit' is dubious at best. Remarkably, in a clear demonstration of the bizarre nature of the assumptions behind these analyses, the Treasury used to count the fuel tax foregone by people transferring from cars to rail as a cost in the benefit–cost ratio assessment. This madness was only stopped in the late 2000s when the then transport secretary Lord Adonis managed to change the rules.

The Omega team stressed the need to move away from the simple 'business case' methodology and to evaluate schemes on a wider basis. Judging schemes solely in business terms, they suggested, was not a proper, balanced way of assessing their level of success; in particular, it failed sufficiently to value the sustainability of projects: 'The perpetuation of restrictive "business case" judgements that essentially de-emphasise "non-business case" considerations and achievements devalues the contribution of planners, project managers and engineers who seek to take a more holistic and long-term approach. This in turn deprives civil society of opportunities to use such projects to transform the economies, territories and cities they serve in line with more sustainable outcomes.'[10] Unfortunately, the pseudo-scientific approach of business cases prevails in all government decision-making on such projects. Politicians like easy numbers and simple concepts they can sell to the electorate, rather than having to put forward complex arguments with caveats.

During the debates over what scheme to develop, another player, in addition to all the various transport bodies, government agencies and local authorities, had entered the field and was quite prepared to throw its weight around. This was the main developers of Docklands, Olympia & York, then owned by the Reichmann brothers, an ambitious trio who had already built schemes in Toronto and New York, and had created what was, for a time, the world's biggest development company. Docklands would, over the ensuing three decades, benefit from three successive rail lines serving its office blocks and each, in turn, was on a far grander scale, both in terms of cost and capacity, than its predecessor. First there was the modest little Docklands Light Railway (DLR), then at the turn of the century the

Jubilee Line Extension and finally Crossrail. And none would have been possible (or necessary) had not the previous one been built. Docklands indeed became a – if not *the* – key determinant of London's transport strategy during this period and played a crucial role in the story of Crossrail by first almost killing it off, and then by ensuring it happened.

The Docklands in east London was a series of ten or so docks mostly built in the nineteenth century to accommodate the huge maritime trade stimulated by Britain's empire. Constructed principally on the north side of the Thames, they dominated the area, being by far the biggest local employer as large communities built up around them. The last to be built was the largest and easternmost – King George V Dock, completed in 1921.

Although greatly damaged by German bombing during the Second World War, when they were a key target, the Docklands were rebuilt and enjoyed a brief resurgence in the 1950s. However, the East London docks were killed off by the advent of containerization, which required much larger ships and consequently far deeper docks, which were duly built at Tilbury, further downriver. The East London docks began to close down one by one in the 1960s and 1970s, eventually leaving a vast derelict site of around eight square miles on the north side of the River Thames. Communities were left to ossify and many buildings lay abandoned, with more than 80,000 jobs disappearing in the 1960s alone.

In light of the massive development today, which continues apace,* the scale of dereliction and abandonment just three dec-

---

* As witnessed by the huge number of partially constructed buildings at the time of writing, in June 2018.

ades ago seems inconceivable. The land was virtually worthless and parts of it were polluted. The catalyst for the creation of today's Docklands was the government-created London Docklands Development Corporation, the brainchild of Michael Heseltine as environment secretary (1979–83), and the first in a series of development corporations established around the country.

The whole concept was a highly political one, an experiment on a grand scale in circumventing the normal planning constraints by effectively shutting out local people from the process and allowing decisions to be made behind closed doors. The new system of development corporations encouraged the market, and thus developers, rather than local councils, to determine how a rundown area would be regenerated.

Launched in 1981, the Development Corporation had an Enterprise Zone that included the site for Canary Wharf, which gave business a ten-year council rates holiday and generous capital allowances. The Development Corporation was additionally able to invest hundreds of millions of pounds of government funding in order to kick-start the process, much of which was eventually spent on transport infrastructure.

Initially, expectations were low and the new buildings were all warehouses and low-rise back offices. They were connected by redbrick roads, clearly designed for light use rather than the hurly-burly of today's Docklands. Apart from subsidizing a few buses, no new public transport system was envisaged or, in fact, felt to be necessary. Then, along came G. Ware Travelstead, a Kentuckian with a massive plan for the area. Docklands would no longer be just another east London suburb with nice views of the Thames, but, rather, a rival to the City of London, able to attract major company headquarters with two massive fifty-storey office blocks on the Isle of Dogs, a U-shaped piece of land bounded on three sides by the Thames and virtually derelict since the closure of the docks. Travelstead was not able to pull the finance together and soon ran out of money, but the concept was picked up by the Reichmanns' company, Olympia & York, which took over the site in 1987.

The Reichmanns' ambitions were far greater than those of their predecessor and they were greatly helped by the fact that they had strong support from Mrs Thatcher, with whom they met several times. They expanded the project, adding extra buildings around what would become Britain's tallest building, the 235-metre-high (771-ft) One Canada Square* (although the second fifty-storey tower was dropped). The Reichmanns envisaged that once the scheme, which was to be built in stages, was completed, 50,000 people would eventually work there, but in fact the total is now double that and still rising fast as more buildings are completed.

Land transport, however, was a barrier to progress, as we have seen, since most freight for the docks had come in and left by water. Those few railways that had been built were closed along with the docks, apart from the little-used North London line which ran to the east of the Isle of Dogs to North Woolwich, an area cut off physically even from the rest of the borough of Newham, let alone other parts of London. The occasional trains were invariably empty.

The vital aspect of transport for the thousands of people

---

* Now in second place as it has been superseded by the Shard at London Bridge.

expected to be employed in the newly expanding Docklands was considered rather belatedly. The Reichmanns originally envisaged that parking spaces would be needed to accommodate the majority of the workers. As Michael Schabas puts it: 'In virtually all megacities, even Paris and New York, most top executives actually drive to work ... it was only after they arrived in London that Olympia & York learned that inner London did not really have an "expressway system"',[11] unlike every US and Canadian city. However, a maximum of just 8,000 spaces was planned for by the Development Corporation. The connection between the City and Docklands was made easier later, when the most expensive road (per mile) ever built in Britain, the £295m Limehouse Link, a tunnel under the dock linking the approach from Tower Bridge with Canary Wharf, was completed in 1993. However improving the roads was never going to be the solution to bring the growing workforce to the Isle of Dogs. A mass transit system using rail was without question the only answer.

Reg Ward, the innovative boss of the Development Corporation, realizing that there was little money available, opted for a light-rail system which would be cheap to build as it could run on some of the alignments abandoned by long-closed railways. Furthermore, the Docklands Light Railway was to be driverless, its cars controlled centrally by a computer. There would, however, be a train captain on board every train to look after passengers and take the controls if the automatic mechanism failed – which in the event proved to be very often.*

---

* Sitting in the front of a DLR train while it bumbles along the rather winding and slow tracks of the system, and pretending that you are at the controls, remains one of London's tourist attractions.

The original idea was that the trains would run alongside traffic on the streets for part of the route, but Ward wanted an automatic control system which could not be used at street level where the trains mixed with vehicles. His favoured automatic system was a far cheaper way of operating the railway and only slightly more expensive in terms of the cost of building it, and the decision to go ahead with it effectively ruled out street-running. Instead the Development Corporation and London Transport devised two lines, within the budget of a mere £77m, which ran largely on the route of abandoned railways. The first went from Tower Gateway, on the southeastern edge of the City, to Island Gardens, at the southernmost river frontage of the Isle of Dogs; the second ran from Canary Wharf north through to Stratford via Poplar. Remarkably, as the latter used old alignments it cost only £10m, which must make it the best-value railway ever built in London. Nevertheless, under the inflexible benefit–cost methodology, neither line was able to show a viable business case. The Development Corporation and London Transport simply went ahead, trusting their instincts that the scheme would work and knowing that without the railways the gleaming new towers of Canary Wharf would be virtually inaccessible to those who worked in them.

By the time of the publication of the *Central Rail Study* in 1989, Docklands was no longer a lost world at the end of the North London line but the site of the world's largest development. And Olympia & York were now serious players, able to throw their weight around and effectively determining London's transport policy. The company was only interested in bringing people to

Docklands and had little understanding of the hostility from some councils and local people to their project. They thought that all they had to do was ask Mrs Thatcher and things would get done.

The DLR was, inevitably, a pretty ramshackle and unreliable system at the beginning. Initially all the trains consisted of a single carriage, which, it was claimed, could carry 260 people, though 160 was a more realistic estimate. With the trains running every four minutes, this gave a theoretical maximum capacity of 4,000 passengers per hour. Together with the parking spaces and buses, this meant the development could possibly cater for 30,000 workers, but in the light of Olympia & York's growing ambitions this was plainly inadequate. The DLR, therefore, became an ever-expanding railway.

Michael Schabas, who worked at various times for both Olympia & York and the Docklands Corporation, argues that 'without the DLR, no Canary Wharf. Without Canary Wharf, London would have grown more slowly as quality office space became scarce. International businesses would have been even more frustrated with London's high costs and some would have relocated to Frankfurt, Paris or perhaps even Reading.'[12] And without Canary Wharf it would have been impossible to justify Crossrail. All this thanks to a modest little railway that initially cost just £77m (about £200m in 2018 money) to build. In many respects, the DLR is the opposite of Crossrail. It is a series of small schemes built up over three decades to create a network that now does much more than simply serving Docklands, though that definitely remains its main task. Crossrail, on the other hand, is a megaproject created in one go as a huge new transport system for London, though, like the DLR, it may well evolve over time by, for example, serving more destinations outside the tunnels than it will when it opens.

As Schabas describes it, the DLR 'looked and sometimes even operated like a modern "mini-metro". But all too often the system simply ground to a halt. "VIP" visitors were sometimes stranded when they tried to reach Canary Wharf.'[13] Schabas gave evidence to a Commons committee 'a bit too honestly', as he puts it, that 'delays experienced by passengers are between 10 and 100 times greater than found on comparable systems'.[14] It was an embarrassment for London Transport and the Development Corporation, and it made Olympia & York – who, after a hiatus, had begun to fill up its huge office blocks – recognize that improving public transport was a priority. First, the company negotiated a deal with London Transport for new two-car trains – the original one-car versions could not be extended for technical reasons and were sold to Essen in Germany – agreeing to pay £68m, about 40 per cent of the cost of the new cars and the related platform extensions. It was, however, never going to be enough, since the employment predictions for Docklands were rising annually, even though a couple of extensions were soon being constructed. One went into the heart of the City, bypassing Tower Gateway and plunging underground in a tunnel to a platform from which passengers could access Tube trains at Bank via a rather circuitous route along narrow passageways. Again Olympia & York provided around 40 per cent of the funding, which, because of the cost of tunnelling, was £276m, more than three times the original cost of the whole DLR. The other extension, for a similar cost, went as far as Beckton and was intended to provide better connections with the easternmost Royal Docks. It has proved be something of a white elephant, although it does now serve the huge ExCeL exhibition and conference centre in Custom House, which opened in 2000.

Olympia & York realized that it could not rely on simply having a single, rather basic, rail link to its massive development and started pressing for a longer-term solution. The DLR, useful as it was, could therefore never be the whole solution and the Reichmann brothers pushed for a Tube line into Canary Wharf. Olympia & York, impatient at the slow progress of London Transport's plans, was busy developing its own plans even as the *Central London Rail Study* was being prepared. The company had initially considered various possible extensions of the Bakerloo Line, either from its terminus at Elephant and Castle or directly from Waterloo, but these ideas were quickly abandoned as they would have involved complex engineering and had damaging environmental implications. Instead, the company seized on an idea from a London Transport document setting out a 'Docklands Public Transport Scheme', published in October 1988. The key recommendation was for an underground line that would provide 'a new South Bank corridor rail line from Waterloo to Canary Wharf, the Blackwall Peninsula and possibly to Stratford or Woolwich which would benefit cross-river access especially to the isolated Rotherhithe Peninsula, the Isle of Dogs and Blackwall Peninsula, and provide an additional link to central London and additional capacity to meet future growth'.[15] Olympia & York seized on the idea of linking Waterloo, where many of the workers at Canary Wharf would arrive from their Surrey and Hampshire homes, and the Isle of Dogs. However, they were not at all interested in serving southeast London and therefore suggested a direct service linking the two, with just one intermediate station at London Bridge. They called this the

Waterloo & Greenwich Railway, a name redolent of some quaint little Victorian branch line or a mere two-station shuttle railway like the Waterloo & City Line. The idea was that it would continue under the Thames to the abandoned British Gas site at North Greenwich, which later became the location of the Millennium Dome (now part of the successful O2 entertainment district).

The Waterloo & Greenwich scheme was worked up quickly by an Olympia & York team led by Schabas, with the co-operation of London Transport, with the expectation that it would eventually become a joint venture. This was a remarkable exercise as the public sector had long been the only source of such major transport plans. It was, indeed, almost a reversion to Victorian times when private investors would put their schemes forward for parliamentary approval. The company even placed their plans for the line out to tender to four bidders who priced it at between £450m and £600m, rather less than the £800m London Transport had estimated. Olympia & York were therefore obliged to commit substantial funding towards its construction.

However, London Transport's reaction to the twenty-page report setting out the plan for the Waterloo & Greenwich was, Schabas says, 'hostile': 'LT was a proud organisation, with a long history of public service. They seemed unable to accept the idea that "foreign developers" would start digging tunnels under London.'[16] London Transport also found the idea of running through southeast London without any intermediate stops completely unacceptable. It was a tenet of transport planners that the Underground should serve the communities through which it passed.

The government, who were intent on keeping Olympia & York onside, commissioned yet another report, the *East London*

Rail Study, even though the *Central London Rail Study* had not yet been published. This time the work would be outsourced to Halcrow Fox Associates, consultants who were independent of London Transport and British Rail. Its report, published in the summer of 1989, recommended a route that corresponded almost exactly with the eventual alignment of the Jubilee Line Extension, with a couple of possible variants. Olympia & York was broadly satisfied, though its executives wondered why there were plans to have stations at Bermondsey and Southwark that would slow the trains down by a couple of minutes and serve areas of largely public housing for the sort of people who were unlikely ever to work at Canary Wharf. A campaign by London Underground's managing director Denis (now Lord) Tunnicliffe, and the prominent local Liberal Democrat MP for Bermondsey and Old Southwark, Simon (now Sir Simon) Hughes, ensured that these stations were built as a key part of the regeneration of this part of southeast London. The company also took issue with various other aspects of the building of the new line; in particular, it wanted to remove any embellishments, such as architect-designed stations, in order to keep the potential cost, and consequently its financial contribution, to a minimum.

Nevertheless, Olympia & York, which had tried unsuccessfully to publish a bill setting out its original plans in 1988 (private bills have to be deposited by 27 November if they are to stand any chance of being passed by both Houses of Parliament in that parliamentary session), agreed in the autumn of 1989 to help finance a second attempt to submit plans to enable the construction of the Jubilee Line Extension jointly with London Underground and London Transport. London Underground, in fact, only agreed to do this through gritted teeth. Its bosses could not forget that the Jubilee Line Extension had come out third best in the *Central London Rail Study* and still wanted to see Crossrail and the Chelsea–Hackney lines built first. However, while the methodology of the *Central London Rail Study* pointed to Crossrail, the politics pointed to the Jubilee Line Extension. The Extension managed to get ahead in the queue, thanks to pressure and detailed work from Olympia & York, who saw it as essential for their huge project and were in effect determining London's transport strategy, with the help of the Conservative government. As Jon Willis, who as a London Transport planner helped produce the *Central London Rail Study*, later wrote in a London Transport pamphlet published in 1997 when work on the Jubilee Line Extension was about half completed:

> Since the time of the *Central London Rail Study* the debate over whether or not the Jubilee line project should take preference over the other central London rail projects has continued. *Other projects would undoubtedly provide higher benefits but the Jubilee line extension would do more to assist new development and encourage regeneration of the large Docklands area.* In the end the decision to go ahead [with the Jubilee Line Extension] was undoubtedly influenced by pressure from the developers of Canary Wharf and their willingness to provide a direct financial contribution [*my italics*].[17]

In other words, the government favoured a project that was designed to help a private developer seeking to create new markets rather than relieve the congestion problems faced by a large number of its existing passengers. Transport Minister Steve Norris admitted then to the Commons' Transport Committee:

> It is occasionally the case that schemes like the Jubilee Line [Extension] are authorised for construction because although it is recognised that they do not meet the standard cover tests, there are nonetheless regeneration issues at large there which demand a call on the public purse. It is unhelpful and would be misleading solely to believe that cost benefit is the only criterion on which investment should be based.[18]

In other words, the government ignored the precise – and supposedly sacrosanct – methodology that was supposed to inform its decisions. Remember that this was the heyday of Thatcherism whose credo was that developments should be market-led and privately financed. The Channel Tunnel, which Thatcher had insisted should be entirely funded by the private sector (notwithstanding some fiddles involving British Rail having to purchase future capacity), was being built, and the so-called private finance initiative (PFI) was being increasingly used. Furthermore, the GLC having just being abolished, London lacked a public body able to determine a long-term strategy (and would do so until the establishment of the Greater London Authority in 2000). Following the abolition of the GLC, London Transport and its subsidiary London Underground were under the control of the Department for Transport and constrained from acting independently.

In the context of the zeitgeist, Olympia & York's ability to influence London's transport strategy was hardly surprising. In the event, little of the financial contribution materialized because Olympia & York went bust just before making the first £40m payment. Although the company's contribution was claimed by the government to be £400m, in effect it was worth less than £150m as it was to be paid for over a twenty-five-year period and therefore its value should have been discounted.* Crossrail, meanwhile, was to be collateral damage of Olympia & York's victory. While its planning would be continued for the time being, it would fall at the first legislative hurdle, putting its future in doubt and eventually resulting in a delay of nearly two decades.

---

* There was later a dispute over the payment because of the poor per-formance of the Jubilee Line and consequently the contribution was further reduced.

4.

Saved but shelved

When the Jubilee Line Extension received the go-ahead there was still the expectation among London Transport's planners that Crossrail would also soon be built. After all, they argued, the business case was good and the need was all too obvious. Following the publication of the *Central London Rail Study* in 1989, a Cross London Rail Links team was created as a joint operation between London Transport and British Rail to examine the concept further.

While openly favouring the Jubilee Line Extension, as Schabas puts it, 'government ministers did not want to be seen to kill off a scheme that was supposedly going to give every London commuter a seat on the morning train'.[1] The scheme envisaged by the *Central London Rail Study* would have connected into the Metropolitan Line through the junction under Hyde Park and thereby relieved pressure for commuters coming from northwest London as well as those on the Central Line. The price London Transport had to pay for keeping all the various schemes on life support was that it was told to choose one line to progress, either the Chelsea–Hackney Line or the east–west Crossrail alignment.

The Cross London Rail Links team considered the two options, on the basis that the Jubilee Line Extension would now definitely be built. On the face of it, Chelsea–Hackney was the best option. Although the team estimated it would cost about 20 per cent more to build than Crossrail, it would bring in about 50 per cent more passenger revenue. The benefit–cost ratios of the two lines were very close – both were around 1.3–1, which is poor – and according to Schabas, who is an expert on business cases, this convergence was not a coincidence but was a way of getting the civil servants off the hook: 'These figures are generated from large "black box" models with thousands of input assumptions, but nobody really believes the results are accurate even to a single decimal place.'[2] He suspects that the planners manipulated the figures to ensure that it would be the politicians, rather than the civil servants, who would have to select which scheme to go with first.

The ploy worked. The decision went right to the top. Neil McDonald, a senior civil servant in the Department of Transport, was summoned to Number 10, where to his surprise Mrs Thatcher kindly poured the poor fellow, who was not used to finding himself in such lofty company, a cup of tea. McDonald set out the case for both lines without making a recommendation. Mrs Thatcher, however, was not fooled. When he had finished she turned to Cecil Parkinson, her transport secretary, exclaiming: 'Hackney! Hackney! Do you know what sort of people live out in Hackney? They are not Conservative voters. Who wants to go to Hackney. They are not our people. No, tell me about Crossrail.'[3] Parkinson had no answer, though he could have pointed out that the voters of Chelsea, who would have benefited most from the line given the paucity of public transport in that rather cut-off part of west London, were not known for their Labour proclivities. Consequently, there was no contest and Crossrail got the nod over Chelsea–Hackney. This meant that the decision to proceed with Crossrail over its rival scheme, now called Crossrail 2, had nothing to do with its advantages such as the potential linkages to Heathrow and Docklands, nor its benefit–cost ratio, but, rather, as ever, naked and raw political calculation. A civil servant involved in the process later said: 'It is a classic example of where a year's work boiled down to a ten-second political prejudice.'[4] London Transport was, therefore, allotted the task of proceeding with the detailed design of the Crossrail scheme and presenting a bill to Parliament by the autumn of 1991.

Numerous options were considered for the sections running at ground level, but the alignment of the core tunnel linking Paddington with Liverpool Street was settled relatively early. Digging deep holes under London is no easy task. As well as obvious obstacles such as sewers, gas and electricity mains and Tube lines, there were other potential constraints. As Schabas says, 'rail planners learn that in densely built-up cities you don't "plan" a route as much as "find" one'.[5] Don Heath, the veteran railway planner who worked on the Crossrail's route under central London, told me that 'there are all sorts of hidden things, and we knew we had to avoid things that are listed but then there are others which do not exist, officially at least'. One day, he was told to expect a call from someone who would not reveal his identity – 'no name, no pack drill'. The man rang Heath at precisely the appointed time, telling him, 'I don't exist but you were expecting a call from me'. After the usual courtesies, Heath reports, 'he told me that "we have studied your proposed alignment and I am pleased to tell you that it does not interfere with any of our assets"'.[6] There was much relief in the Crossrail team. Amazingly, the alignment of the tunnel survived to become pretty much the route that has now been built, described fully in Chapter 6, though there were subsequent minor adjustments as the detail was developed once the line was given the definite go-ahead. Moreover, some extra stations have been added in response to pressure from local interests and developers, plus a southeastern extension added, partly in tunnels.

The plans for the western section were, however, radically changed. The planners realized that there were major potential complications on the Metropolitan Line branch where power is taken from a third rail rather than from overhead line equipment. Either the trains would have had to be adapted to give them the ability to pick up electricity from a third rail as well as from overhead wires, or wiring would have to be constructed along this section of the Metropolitan. Although it is technically possible to do this, it can cause complications with earthing arrangements while using two different sources of power would have imposed considerable extra costs on the rolling stock. A further problem is the incompatibility of the signalling systems as the emergency stop systems that activate when a signal is accidentally passed at danger are different on Underground and mainline trains. These complexities are commonplace as the rail network was built by a variety of Victorian entrepreneurs and upgraded with different equipment over more than a century, and represent one of the difficulties facing rail planners drawing up new schemes who have to take into account the considerable extra costs incurred.

Those commentators who are quick to pen angry articles berating the rail industry for its shortcomings would do well to bear in mind the sheer complexity of the technical issues with which planners and engineers have to contend.

Two key decisions which the Cross London Rail Links team made at this early stage have survived right through to the completed scheme. First, they decided that the tunnels would be 6 metres (19.6 ft) in diameter (later expanded to 6.2 metres to allow for an emergency walkway along their entire length), which would permit the passage of mainline trains rather than only Tube-size ones. Richard Malins, one of the authors of the *Central London Rail Study*, told me, 'The Railway Inspectorate would not have liked the idea of light Tube trains running on the main line, even though the legacy ones, such as on the Bakerloo, are allowed.'[7]

The second decision was to have very long trains, capable of carrying 1,000 or more people and which consequently were ten or even possibly twelve cars long, requiring platforms to be 240 metres (not far short of 800 ft) – nearly a quarter of a kilometre – in length. The intention was to make sure that Crossrail would have the capacity to cope with decades of growth in passenger numbers, but the corollary was that the underground stations – always the most expensive part of a project, because of the complexities associated with building them – would have to be very large. This also determined a key aspect of the design of the stations, a decision which, according to Meads, was made very early in drawing up the scheme: 'All the stations in the tunnel would have two entrances, which greatly increases the potential catchment area and consequently will attract more people to use the railway.'[8]

Despite the many decisions that had to be made and the complexity of the issues, the planning process was completed relatively quickly and in November 1991 a private bill was submitted to Parliament by London Underground and British Rail with a cost estimate of just over £2bn at 1993 prices (say around £4bn in 2018 money). And that's when the trouble started. Already there were signs of half-heartedness on the government side. London Transport had wanted a hybrid bill, one in which both private and public sector matters are taken into account, but the government had refused because it would then have been directly involved and consequently any failure would have reflected on ministers.

To make matters more difficult, the country's economy was beginning to deteriorate. Even as the bill was being finalized and then deposited, its very *raison d'être* was being undermined. A recession, caused by an oil price rise and the overheated economy, had started a year before the bill was submitted and would last right through to the end of 1993, reducing passenger numbers on the London Underground and the South East's suburban rail systems.

The procedure for private bills involves the creation of a special Commons committee to scrutinize the project. Opposed Bill Committees, as they are called, have a quasi-judicial status as they are effectively granting planning permission to the scheme under discussion. The process can be both onerous and lengthy, and MPs are often reluctant to sit on them. Members whose constituencies would be impacted directly by the legislation are barred from being involved. Therefore, the process of selecting members can be arbitrary and it is not unknown for whips making the selection to inveigle miscreant MPs, who may in the past have disobeyed the party whip, into sitting on them. Although the procedure is supposed to be independent of government, the whips may select particular MPs to sit on the committee to ensure that the bill gets through or, more likely, to kill it off. The chairman of the committee that considered the first Crossrail bill, Tony Marlow, MP for Northampton North, certainly fitted into the second of those categories. A right-winger and hard-line Eurosceptic (he would later lose the party whip as a result of his opposition to the 1992 Maastricht Treaty), Marlow was instinctively against big-project spending and may well have been chosen to reflect Treasury scepticism about the project.

However haphazard the method of choosing its members, the Opposed Bill Committee rather unexpectedly approached its role as arbiter of the Crossrail scheme very seriously. As one witness put it: 'This was Parliament at its best. They looked at everything from finance to the London job market, the effect on existing train services and the environmental benefits.'[9] Their deliberations were certainly onerous and complex, the 1991 Crossrail Bill Committee sitting for as many as thirty-two sessions.

All the while, the Treasury was working beneath the radar to delay, or, better still, to kill off the scheme. All its most Machiavellian tactics were being deployed. At every public spending review, moneys allocated for transport were being reduced. The Department for Transport's investment budget tends to be low-priority which meant it is always an easy target because of the long-term nature of transport schemes and Crossrail was, in any case, constantly in the Treasury's crosshairs. Apart from direct cuts to proposed transport budgets, which put pressure on the Department to rein back on any major schemes, there was a constant round of reviews of the Crossrail project, focused in particular on the potential for private sector involvement. This had become a 'must have' rather than, as before, 'a nice to have' for any such scheme, and, since there was no great appetite in the private sector for investing in transport, this requirement had the effect of delaying or even killing off proposed schemes. The series of reviews of the Crossrail project revealed little that was new but put pressure on the promoters of Crossrail to justify the project. One report, by Scott Wilson Kirkpatrick, found that the scheme had 'no "fatal flaw", was well-conceived and would be an attractive project for the private sector'[10] while another, by SG Warburg, optimistically claimed that 40 per cent of the capital cost could be privately financed. A third concluded, with a triteness that would have embarrassed an A-level student, that Crossrail 'attains a range of benefits going beyond its stated purpose as a transport function'.[11] The Treasury nevertheless went on to commission yet another study by a group of consultants which included Bovis, the project management and construction specialists, and Schroders. At last, this report seemed to give the moneymen the answer they were seeking: it called for the project to be delayed since 'it was a visionary project which may be ahead of its time in view of the change in employment and commuting patterns'.[12] As one planner involved in the project put it to me, 'I said at the time that the only reason they would stop doing studies is if they ran out of consultants to do them or if they finally got the right answer.'[13]

Both politics and economics were moving against Crossrail. The emphasis on private funding was perfect for the Treasury. Not only did it lead to delays, since the source of such funding was unclear – there was no rush of private investors coming forward with offers of money – but it was also in tune with their agenda. After John Major unexpectedly won the 1992 election with the manifesto commitment of privatizing the railways, British Rail was increasingly unable to commit any funds to the project since it was about to be broken up. Although initially Railtrack, which took over the track and infrastructure in the break-up of British Rail, was going to be kept in the public sector, there were moves immediately following the passage of the Railways Act 1993 to put it up for sale, and having a commitment to a huge and risky project like Crossrail on its books would undoubtedly have reduced its value.

The numerous studies eventually did the Treasury's dirty work for them. They delayed the progress of the bill sufficiently for circumstances to change as the economy deteriorated. Crossrail was in fact caught up in the battles within the Conservative Party between Major and his supporters and the right-wing Eurosceptic fiscally conservative 'bastards', as Major called them, who were very much in control of the Treasury.

The underhand methods employed by the Treasury, rather than a full-on attack, were necessary because of the public enthusiasm for the project expressed by John Major and some of his senior colleagues. Lobbying by Crossrail's supporters, which attracted favourable publicity for the scheme and support from business and the City of London, kept the project alive for far longer than might have been expected given the strength of opposition and the tanking of the economy. While the Treasury huffed and puffed behind the scenes and tried to undermine Crossrail's progress, the main political players seemed, at least superficially, to be trying to ensure that it would be approved. At the Tory Party conference in October 1990, the transport secretary Cecil Parkinson had announced that the scheme would go ahead at a cost of £1.4bn, with work starting in 1993 and with a completion date of 1999. John Major, who replaced Mrs Thatcher as prime minister a matter of weeks after Parkinson's conference speech, was also, in public, keen to see the scheme go ahead (even though he was a former chancellor of the exchequer). In May 1993 he even stated that the bill would receive parliamentary approval within a month. Labour, too, were generally supportive and it therefore seemed likely that the scheme would indeed proceed.

It was not to be. The bill had been introduced into Parliament in November 1991 but did not receive its second reading for another eighteen months as it had no real champion. Unusually, private bills do not necessarily fall when an election is called, but after the Tories' unexpected victory in April 1992 Malcolm Rifkind, who had strongly supported Crossrail, was replaced as transport secretary by the far less enthusiastic John MacGregor. The bill finally received its second reading in May 1993 as the Department for Transport wanted to ensure that the Jubilee Line Extension had safely passed through Parliament before Crossrail was debated.

It was then that the going got tough. The bill attracted 314 petitions from opponents or people desiring to make changes, which were considered by the committee of four MPs chaired by Tony Marlow over a seven-month period in 1993–4. There were the obvious Nimby arguments from people directly affected by the construction work, ranging from posh Mayfair residents to rather less affluent Bangladeshis in Tower Hamlets. There were, also, mysterious objections from US Embassy officials who told the planners that their government wanted to stop the construction. Despite the assurances made to Don Heath that Crossrail's alignment kept clear of any secret structures, the Americans were apparently concerned about the effect of the passing trains, equipped as they would be with a complex range of electrical equipment, on a mysterious building apparently owned by the US Navy near their embassy in Grosvenor Square. The US government never elaborated on why it was objecting, and only the most imaginative conspiracy theorist would suggest that it was anything but a coincidence that the Crossrail tunnels will now open two years after the US embassy moved south of the river to its monstrous new 'bunker in the air' at Nine Elms in January 2018.

Leo Walters, a surveyor who ran the Mayfair & St James's Residents Society and who was the Nimby-in-chief, argued that the alignment was wrong and suggested instead that the route should follow Marylebone Road and connect with King's Cross St Pancras. Other objectors took the same view. Ken Purchase, one of the Labour MPs on the committee, later told me: 'The connection with the Channel Tunnel Rail Link is very important… Under the present Bill, it would not have been possible to build a link later because of engineering difficulties.'[14] This was wrong on both counts: the connection with the Channel Tunnel Rail Link would have provided very few extra passengers and a later link could easily have been built (though the business case for it would have been weak). Indeed, London Transport countered the argument for a connection with King's Cross St Pancras by stressing it was already the best-connected point on London's rail network, with five Underground lines and a range of suburban services including Thameslink. The real need, they said, was to make the West End more accessible.

Walters and his fellow travellers were appalled that anyone could consider it a good idea to run a railway under Mayfair

– the location, after all, of the most expensive property on the Monopoly board. He would later tell the Crossrail committee: 'You can imagine if you, yourself, sir, were living where we live in Mayfair how you would feel. I am not an expert but these tunnels are eight metres [actually 6.2 metres] in width and all that noise and the rest of it… this ginormous project is being built in the wrong place.'[15] The truth was that the tunnels would be dug deep under London and would have little impact on the residents (bar the inevitable holes in the ground for station access and ventilation shafts). But this did not stop these protesters, 'particularly all the posh folk in Belgravia and Mayfair'[16] (as one of Crossrail's political lobbyists described them), from 'screaming'. A new railway line passing under the front doors of just about every influential interest group in the country was always going to cause consternation, and it is testimony to the advocacy and PR skills of the later Crossrail promoters that they eventually managed to get the scheme built.

While the Mayfair lot may have been far more powerful, it was, surprisingly, the rather less affluent residents of Tower Hamlets – at the other end of the tunnel – who inflicted more lasting damage on the Crossrail scheme. They were more effective partly because their case was better, since they would have to endure the permanent disruption caused by Crossrail's eastern portal emerging in a rare green space, but more specifically because the local council employed the experienced transport consultant Jim Steer, one of the founders of a major consultancy, Steer Davies Gleave, to spearhead its opposition to the scheme. Steer, while being generally supportive of the Crossrail concept, had specific objections which he felt had not been addressed by the promoters: 'The focus of the scheme was entirely on central

London, which was the result of the fact that it had emerged from the *Central London Rail Study*. There were five stations being built in central London but there was nothing for inner London, no station between Liverpool Street and Stratford, and the first stop after Paddington was Acton on the main line and Wembley Park on the Aylesbury branch.'[17]

In Tower Hamlets, there was widespread concern – particularly among the local Asian population – about the dust and disruption likely to be created by building work in the area. Understandably, too, local politicians argued that there was nothing in the scheme for the borough, apart from hassle, dirt and noise. Worse, the Crossrail team had unwisely selected Allen Gardens as its eastern portal, one of the few substantial green spaces in the borough, which would be completely destroyed as a result. The Treasury, spotting an unexpected ally, had several secret meetings with Tower Hamlets officials and councillors.

Steer was doubtful about the underlying case for the project: 'London Transport were basically saying that they could not build more roads, so railways were undoubtedly the answer – and Crossrail was the right scheme. It was not a convincing case.'[18] Steer faced seven gruelling days of questioning by the committee and maintains to this day that he got the right result even though now he is a fervent supporter of Crossrail: 'At the time, it was just not the right scheme. It did not serve either Heathrow or Docklands, and did nothing for inner London. The London Transport scheme made comparisons with the French RER but that was not valid. The RER is part of a wider strategy of linking the suburbs with central Paris, whereas there was no strategic thinking behind Crossrail. The very fact that it concentrated so much on "central" London demonstrated that, whereas the RER was conceived to ensure the people in the *banlieue* had better access to the centre.'

Steer reckons that his small team of just two or three people demolished the case for the line despite being up against serried ranks of experts and lawyers. He was helped, however, by having on his side one of the most eminent planning lawyers in the country. George Laurence QC, Steer explained, 'gradually teased out the fact that Crossrail, as conceived, was the wrong answer'.[19] As Steer told me, 'the lack of a strategic transport plan for London meant that hardly any scheme makes sense on its own. For example, as a way of getting people out of their cars and onto the railway, Crossrail is very expensive and pretty ineffective.'[20] London desperately lacked an organization to co-ordinate its transport policy following the abolition of the GLC and that's why, Steer believes, the project did not add up.

Steer also questioned the funding of the scheme – and here the promoters were on particularly thin ice. The government's emphasis on ensuring there was considerable private-sector involvement meant there was no clear commitment from ministers that the scheme would go ahead. And there was simply not sufficient interest among private investors to obtain the 50 per cent share – which the government was seeking – of what was now estimated as a £2bn scheme. London Transport had made vague statements about developing the sites above stations but there was no detailed explanation of how this might work. Steer says: 'If you do not have the Treasury supporting a scheme, and the Treasury was adamant it was only ever going to put in 50 per cent, then you have one hand behind your back.' Steer was unimpressed by the financial information being put out by the Crossrail's promoters: 'We asked about funding, and they finally produced a paper, but you have never seen such a heavily redacted document – it was short and half of it was blanked out.'[21] As I wrote in a commentary in *Independent London* when the scheme was 'nixed', as the headline writer put it, 'Crossrail was originally conceived as a public sector project but the Government obsession with the private sector means that it now wants to transfer some of the risk to the private sector... The problem is that involving the private sector will mean the taxpayer getting a worse deal than if the project was entirely publicly funded... the private sector will have to be rewarded for its daring with a healthy income stream from the line, much higher than the return expected by the Treasury.'[22] Interestingly, the lesson was learned, as there was barely an attempt to obtain private sector funding for the revived scheme in the early 2000s.

Steer's evidence was clearly damaging because it came from an expert source. But it was the loss of passengers caused by the downturn in the economy, which did not start to recover until 1994, that fatally undermined the case for the project. There were, too, some rather more substantial questions about the economic case, which indeed was weak according to the very methodology used by the planners. The fundamental problem was that public transport usage, far from growing, as predictions had suggested, was actually decreasing, and, inevitably, costs were going up. The benefit–cost ratio was worsening daily. The overcrowding that had been at the root of the analysis in the *Central London Rail Study*, had been greatly reduced, with passenger numbers on the Underground down by 20 per cent since 1988. The argument about congestion relief could no longer be sustained to justify the scheme. London Transport tried to keep the momentum going by suggesting – rightly, as it turned out – that the economy would eventually start growing again in line with the usual economic cycle. However, Steve Norris, the transport minister charged with pushing through the scheme and a great supporter of the project, certainly found it a struggle in a debate on Crossrail in June 1993:

> It is of course true that the employment levels on which the plans for Crossrail were originally based – the assumptions in the *Central London Rail Study* – are now some five years old. Of course, they look optimistic now because they were made before the depth of the recession became apparent. Traffic levels have certainly fallen since the assumptions were made. It may well take longer to achieve the forecast traffic levels, but we are confident that, with economic recovery, those levels will be achieved, albeit later than the central London rail study supposed. I underline that Crossrail is not a five-year or ten-year project. As with all the great underground rail projects in London, it is a project for the next 50 or 100 years.[23]

These arguments were torn apart by the four members of the committee charged with the arduous task of examining the scheme and listening to the long list of petitioners. They were not impressed by London Transport's long-term view. Witnesses later reported that they seemed intent on not allowing the bill through. The bill was formally killed on 11 May 1994 when the committee voted 3–1 against it, Marlow siding with the two Labour MPs against Crossrail's sole supporter, Matthew Banks, Tory MP for Southport. Marlow did not give any reasons publicly but in essence the preamble of the bill, the statement setting out its objectives, had not been proved. The Labour members,

Ken Purchase and John Marek, while supporting the general concept, seemed to have been uneasy about the proposed financing and, in particular, they were wary of the involvement of the private sector which had been mandated by the Treasury.

While John Major, Steve Norris and the other key supporters licked their wounds, one cabinet minister could not hide his delight at the demise of the scheme as he had opposed it all along: 'Crossrail was ill-thought out. It went from nowhere to nowhere. No sane person has ever wanted to go to Shenfield.'[24] This was to miss the point. The project was not about getting people to Shenfield – which just happened to be a convenient place to turn the trains around – but about lightening the load on London's railway network.

The scheme's supporters did not give up without a fight. There remained considerable support for Crossrail among MPs and, over a period of several weeks, attempts were made to get the bill back in Parliament, and to force the committee to publish the reasons behind the decision. However, even a petition by 280 MPs was unable to persuade the committee to look again at the issue, nor even to come to the House to explain why they had decided to block the bill. A couple of months later, the government moved to push the bill forward – for a scheme that was now estimated at £2.6bn – through a new procedure to get around the stubborn committee, but to no avail. In terms of national transport policy, the government's principal focus was now on rail privatization, which was highly controversial and unpopular, even among some Tory MPs.

As for Crossrail, it was either in cold storage – or possibly even worse. However, a key decision by London Transport would ensure it lived to fight another day.

5.

Crossrail revived

Crossrail could have been dead and buried at this point. It had few friends and plenty of enemies, and the recession had pulled the rug from under its business case. Although the prospect of the line ever being built looked bleak, London Transport kept the faith and made the crucial decision to safeguard the line without which the opportunity to build it would have been lost forever. This move prevented any development being built on the route that would make it impossible for the line to be completed. This not only had the effect of stopping developments that would have jeopardized a future scheme, which some landowners later complained about, but also effectively meant that any revived Crossrail would have to use the same alignment under London.

London Transport retained its optimism, which is why my article in the *Independent London* on the defeat of the bill was subtitled 'Back to the drawing board'.[1] I quoted Jim Steer saying that 'far from being a wasted opportunity for London as business leaders said, it was a huge opportunity to sit back and draw up a plan for what infrastructure London needs'. Essentially, Steer was arguing for a London strategy: 'The committee considered all kinds of issue such as the regeneration of the East Thames corridor, air pollution, employment in London and so on. The issues are bigger than transport and have to be looked at properly, despite the Government's hatred of planning.'[2] There was an expectation that eventually a way would be found to get Crossrail built but even the pessimists among the scheme's supporters must have hoped it would take less than a quarter of a century for the line to open.

The failed Crossrail scheme had racked up spending of £157m and London Transport was keen to ensure that this size-able investment would not be wasted. London Transport there-fore retained a team to continue producing details of the route and to maintain the protection afforded by safeguarding. The Tory government, anxious not to alienate business and the City, both of which were deeply disappointed by the loss of the bill, did not 'have the guts to kill it off'[3] in the words of one of the planners. Therefore the scheme was put on life support.

In July 1994, the government announced that the Crossrail project would be brought forward under the new Transport and Works Act system, which was envisaged as the future way to take major infrastructure projects through Parliament. How-ever, nothing could happen for the moment as the government commissioned a further study to determine whether a cheaper scheme might achieve the same benefits as Crossrail. After a few months it was announced, predictably, that none could be found. There now followed yet another review, in which a different team of officials assessed again whether the funding could come entirely from the private sector, as it had for the

Channel Tunnel, which had just been built. They concluded, not surprisingly, that this would not be possible. The Tories, aware they were probably heading out of office at the forthcoming 1997 election, made no formal responses to any of these reports and eventually told London Transport and British Rail, the joint promoters of the scheme, to suspend their Transport and Works Act application. The Department for Transport already had its hands full with sorting out the Jubilee Line Extension, Thameslink 2000 (ironically, despite its name, due to be completed in 2019, when Crossrail fully opens), the Channel Tunnel Rail Link and the hugely complex process of privatizing the railways.

Over the next year or so, the Crossrail team was wound down and staff were gradually let go. Don Heath is very proud of the fact that he found jobs for all bar one of the 320 employees who were 'let go'. The intervening couple of years had not been wasted. The Crossrail team, still funded by British Rail and the Department of Transport, had refined the route and had also outlined a wide variety of potential services that could use the main tunnel, including innovative routes like Southend to Milton Keynes. Such plans, however, were put in abeyance after the collapse of the bill and the project retained only a skeleton staff of half a dozen people to manage the safeguarding process.

At this stage, even the Underground rather lost interest in Crossrail as its managers concentrated on investing in improving the existing network, whose inadequacies and lack of investment had been all too painfully exposed by the 1987 King's Cross fire. Denis Tunnicliffe, the managing director of the London Underground, had developed the concept of a 'decently modern metro', which would need investment of £1bn per year for a decade. He was also focused on ensuring that the Jubilee Line Extension,

by then under construction, was completed in time for the millennium celebrations.

The tiny Crossrail taskforce was kept on for the remainder
of the 1990s. David Warren, who was seconded to the group
from London Transport and worked on the scheme on and
off for a couple of decades, says that during this period 'there
was considerable work in ensuring that the safeguarded route
remained protected and there was always the expectation that
at some point the scheme would be revived'.[4] Warren was right,
but the revival would take some time.

In 1999, two years after the Labour Party's 1997 landslide
election victory, the deputy prime minister John Prescott, who
was responsible for a mega-department that included transport,
expressed an interest in the project. As Labour had begun to find
its feet, the climate was changing, both economically and politically. London commuting was once again growing fast and two
wider political developments helped put Crossrail back on the
agenda. In opposition, Labour had intimated it would reverse rail
privatization, but once in office it decided instead that it would
try to control it through the creation of a Strategic Rail Authority
to give some overall direction to the industry and to look at its
long-term needs. A second institutional change, which would
prove more enduring and ultimately more significant, was the
creation of a directly elected London mayoralty. Astonishingly,
since the abolition of the GLC in 1986, London, uniquely across
the world for such a large city, had had no overall body to look
after the interests of the capital as a whole. Labour filled this gap
in city government by creating, following a referendum, the office
of London mayor whose role encompasses transport policy,
policing, large-scale planning and various other capital-wide

issues. Even before the first mayor, Ken Livingstone, was elected in May 2000, Prescott, whose passion was transport, had asked the shadow Strategic Rail Authority (SRA) to produce a 'London East-West study' to look at the potential for new rail links.

It was, in fact, one of those studies that was expected to give the answer the politicians wanted. Prescott liked the idea of Crossrail and would include it in his ten-year transport plan, *Transport 2010*, published in the summer of 2000. Most of its schemes, including two dozen proposed light-rail projects, never saw the light of day, not least because of an underlying assumption that the private sector would come forward to finance them. Indeed, *Transport 2010* assumed that Crossrail would be financed by the private sector, despite the previous government's analysis that this was not possible. The public sector's role was limited to contributing £3.5bn for extra infrastructure to the existing network and for rolling stock.

The SRA, still in shadow form as the legislation to establish it had not yet been passed, was Prescott's creation, and it was no surprise that its first major report fully backed Crossrail. Its chairman, Sir Alastair Morton, had a penchant for big projects, having been chief executive of the company that built the Channel Tunnel.

Published in November 2000, the *London East-West Study* made several recommendations for improving London's rail infrastructure (including a notable emphasis on freight, which is normally neglected in such reports). It analysed the respective merits of three potential cross-London railways: Paddington–Liverpool Street, and two versions of the Chelsea–Hackney Line which had now been extended to become Wimbledon–Liverpool Street or Wimbledon–Hackney.

Miraculously, according to the report, the benefit–cost ratios for all three were now all remarkably positive. Based on a construction cost of £2.8bn, the Paddington–Liverpool Street route scored 3.2, by far the best of the three schemes, and consequently the report said it should be prioritized. These positive numbers were helped by the fact that the definition of benefits had been extended. In the words of the *East-West Study*: 'The benefits include the direct and associated benefits and include time saving for existing users, reduced congestion on trains, revenues from generated travel, relief of road congestion and reduction in road accidents.'[5] Although Wimbledon–Hackney scored lower, at 2.1, the SRA said it should definitely also be built, with a time lag of three years behind the Paddington–Liverpool Street plan.

However, while the SRA backed the concept of the tunnel from Paddington to east of Liverpool Street, its scheme differed in several ways from the previously rejected plan. The SRA not only dismissed the idea of linking up with the Metropolitan Line through Old Oak Common but was also lukewarm about the idea of a connection with Heathrow Airport: 'There are issues as to whether a cross London tunnel link is appropriate for an airport link and whether such a connection optimises the use of the limited rail facilities at the Airport.'[6] The study went on to say that demand for connections directly into the West End and the City from the airport was likely to be low: 'A significant proportion of these [air travellers] is heading for London, mainly to the hotel districts in Westminster, Camden and Chelsea.'[7] Instead, it suggested there should be services to the airport from Paddington which also served local destinations, unlike Heathrow Express which ran directly to the airport with a very high fare. There was, too, at the time, a plan to link Heathrow with a service along the North London line to St Pancras, but this never materialized. A service following the Heathrow Express route but stopping at intermediate stations would be introduced in 2005 as Heathrow Connect.

The *London East-West Study* reiterated the arguments in favour of the Crossrail scheme but with the added bonus that passenger and population figures were now moving in the right direction. London was growing again, employment in the centre was booming and the number of commuters was soaring to record levels. 'The current system is near to capacity in terms of the National Rail Network, the central London termini and the Underground. This creates a fragile system where the slightest problem can have dramatic knock on effects across the network.' Peak time travel had increased in the past five years and was expected to keep on growing; without extra rail capacity, London's growth would be constrained. Since, as the report pointed out, it was government policy to encourage people to travel by rail rather than in cars, the report argued that bigticket solutions like Crossrail were the only answer – all music to Prescott's ears.

The report also strongly favoured a hybrid bill process for getting the scheme through Parliament, rather than a Transport and Works Act, because the government would then be tied in to the scheme's progress and ultimate fate. A hybrid bill is steered through Parliament by both government and the private promoter, and this government support makes success more likely. While simpler than the process for a Transport and Works bill, pushing a hybrid bill through Parliament is still a lengthy and convoluted procedure as it is the way planning permission for a scheme is obtained. After the bill is introduced into

Parliament, accompanied by a detailed environmental statement, a committee of MPs is selected to hear petitions against particular aspects of the plan from objectors. This can take a long time and, as a consequence, the Crossrail bill would ultimately take three years to get through Parliament.

Crossrail may have been back on the political agenda, but for such a huge project the road ahead was inevitably rocky: it needed political backing, funding, a champion and a slice of good luck. The scheme would eventually get all of these – but not immediately. The government's reaction to the *London East-West Study* was favourable. The two newly created bodies, the SRA and Transport for London (TfL) – the latter having come into being, via the Greater London Authority Act, as the mayor's transport arm – were told by the government to set up a joint project to develop the scheme. They were provided with a reasonably generous grant of £154m to bring the bill to Parliament.

The core aim of the scheme was to help London commuters, rather than supporting longer-distance travel through London from the regions. This was now established as Crossrail's key *raison d'être*, whereas the Thameslink Programme,* which was being developed at the same time, was seen as linking towns on either side of London as distant as Peterborough and Horsham and Cambridge and Brighton. The upgrade to Thameslink, which is coming on stream in phases, simultaneously with Crossrail, is therefore based on a very different concept. As Howard Smith, Crossrail's operating director, explained,

---

* The Thameslink Programme had already gone past its original target date, hence the dropping of '2000' from its original title.

Crossrail is philosophically different from Thameslink. Crossrail is a long tunnel with services that go a sensible distance into the suburbs. Thameslink, on the other hand, is a national railway pushed through a tunnel in its central section. It is sub surface whereas Crossrail is much deeper down with tunnels that were bored. There are platform doors, with a mix of seating including longitudinal. The line out to Shenfield is very much a metro concept, and so is Maidenhead – and while Reading stretches the point, the trains will go on the slow lines stopping at lots of stations.[8]

Furthermore, Thameslink is part of the national rail network and consequently franchised out with the private operator receiving the fares, whereas Crossrail is owned by Transport for London and will be operated under a management contract, with fares going to TfL.

The SRA and TfL spent the next eighteen months consulting on the routes and preparing the business case on which the future of the scheme rested. Tackling overcrowding on the existing network remained the core justification for the scheme: 'Despite planned increases in capacity on the National Rail and London Underground networks, the overall rail network is forecast to be more crowded in 2016 than at present. In addition there are no plans to increase the highway capacity in central London, so growth will have to be accommodated by public transport.'[9] The Business Case quoted the London plan, which expected to see almost three-quarters of a million more people living in London by 2016, and 636,000 new jobs, most of which would be in the City, the West End, east London and Docklands. Crossrail, it was argued, would improve services for passengers, 90 per cent of whom were travelling in cramped conditions at peak times, in several ways: by reducing crowding on several heavily loaded routes; by increasing rail capacity – by some 7 per cent overall* and 20 per cent for some key flows; and by giving much better accessibility to the West End and the City for large numbers of residents in both east and west London.

However, the Crossrail board (officially known as the Cross London Rail Links board, a joint venture between TfL and the Department of Transport) had expanded the aims for the line far beyond the relief of overcrowding at the mainline stations and on parts of the Tube network. The board recognized that London had grown in importance relative to the regions over the decade since the first plan had been rejected in 1994: 'The original concept was focused on central London's problems but now the consideration is necessarily wider. Central London's economy is of great significance nationally and extending its already substantial job catchment area and improving the efficiency and dependability of access journeys (from home, for commuters and from the international and national gateways for business travellers) is a very important feature of Crossrail.'[10] The board's *Business Case* thus made it clear that the new railway would also support the Labour government's transport objectives in relation to the ten-year plan, stimulate the development of the financial industry in the City and Docklands, create better links with Heathrow and facilitate the regeneration of the Lea Valley and the East Thames corridor.

---

* The figure of 10 per cent is often quoted, too, and that the variation can be accounted for by the methodology as overall rail capacity is a difficult concept to assess.

The greater scope of the Crossrail project, with this wider remit, was the underlying reason for changing the proposed route. While the central tunnel, safeguarded and very difficult to move because of potential obstacles, remained the same, a link to Heathrow was added as 'a clear feature in the new proposal'. But there was still much agonizing about what to do in the west. The idea of running on Metropolitan Line tracks out to Aylesbury was dropped because it was found to have 'a number of disadvantages'. Since numbers commuting into Paddington were insufficient to justify new services, and the Heathrow branch would not require the expected twenty-four trains per hour going through the central tunnel, the promoters suggested a line heading southwest to Kingston to take up some of the spare capacity. It would, the report suggested, allow a link, through a tunnel, with the District and Piccadilly lines at Turnham Green or Chiswick Park, and then continue through 'to serve commuting flows – Chiswick, Kew, Richmond, Twickenham, Teddington and Kingston'. This would provide congestion relief on both national rail trains to Waterloo and Underground services through Earl's Court. The idea of running through on the Great Western slow lines out to Maidenhead and Reading was shelved.

In the east, a station at Whitechapel would connect with both the District Line and the East London Line (which is now part of London Overground). This was, in part, a way of buying off Tower Hamlets council and its residents: the connection would 'significantly broaden the spread of Crossrail benefits as well as serving an area in need of regeneration'.[11] More significantly, the promoters had belatedly recognized the importance of Docklands. They envisaged a line that diverged just east of Whitechapel and continued in a tunnel under the Isle of Dogs, with a station at Canary Wharf, to emerge at Custom House in the Royal Docks. Those would be the only stations north of the Thames before the line plunged under the river to serve Abbey Wood (for Thamesmead) and several other stations on the surface before terminating at Ebbsfleet, where it would connect with the Channel Tunnel Rail Link. There was, at the time, no plan for a station at Woolwich, which was later to be the subject of a fierce lobbying campaign.

The search for the best routes for Crossrail was complicated by a factor that had started to emerge in 1994 – namely the marked increase in the number of trains being run for commuters to meet the rise in passenger numbers. This meant that the tracks were being used more intensively and, together with the fact that operators on the now privatized railway had fixed contracts to run specified services, offered less choice and flexibility in terms of selecting routes for Crossrail services. Besides this, the whole planning process, with more 'stakeholders' involved, was now very much more complicated, as the *Business Case* warned: 'previously it was possible to envisage incorporation of mainline services with little change to the national rail network. The intensification of use of the network in the intervening years means this is no longer possible.'[12] The report went on to say that any solution must avoid creating a situation where the Exchequer would have to pay massive amounts of compensation to franchisees affected by the introduction of Crossrail services. In short, despite the positive mood music surrounding the scheme in the early 2000s, changes in the intervening decade as a result of privatization and the boom in passenger numbers threatened to make the project harder than ever to push through.

Of the twenty-four trains per hour passing through the tunnel, the *Business Case* proposed that, in the west, half would go down the southwest branch, while six trains per hour would serve Heathrow – the implication being that the remaining six would simply turn around at Paddington. In the east, the services would be equally divided between Abbey Wood and Shenfield and the latter, thanks to the retention of some existing services, would enjoy an impressive eighteen trains per hour in the peak. Overall, there would be a 40 per cent increase in services from Liverpool Street. The Crossrail trains were envisaged to consist of ten cars and would be 200 metres (656 ft) long.

Then came the bad news. The capital cost envisaged by the *Business Case* was far higher than any previous estimates, at just under £7bn at 2002 prices. Taking inflation into consideration, that would give a final bill on completion of £7.6bn. The bulk of the expenditure of £7bn was the £4.8bn for the main tunnel section from Westbourne Park, the western portal, to the Isle of Dogs and Stratford in the east. The higher costs were put down to the detailed analysis of the scheme which produced a more realistic overall financial picture than previous estimates, and some project creep, such as the extra branches. The benefit–cost ratio on the benchmark assumptions – numerous alternative scenarios were put forward – was 1.99–1, a suspiciously precise estimate which seemed to suggest that its authors were reluctant to give it their full support since 2–1 would have sounded so much better.

This *Business Case* was submitted in July 2003 to the transport secretary, Alistair Darling, a man who was never impressed by big projects – as demonstrated by his steadfast refusal ever to countenance any work being carried out on High Speed 2 during his four years tenure of the transport brief* and who, indeed, was averse to spending any money at all (as befitted his later appointment as chancellor of the exchequer). His response to the publication of the *Business Case* was suitably mealy-mouthed: yes, this was a great scheme and, yes, it would do a lot for London but, well, it does cost an awful lot, doesn't it? The Department for Transport agreed that Crossrail would 'facilitate the achievement of planned growth targets for London, assist the economic development of London and contribute to regeneration policy' and that it would 'receive widespread support from the travelling public, residents and businesses'.[13] The Department also agreed that access to Heathrow and Stansted airports, both of which were slated for expansion in the recently published Aviation White Paper, would be facilitated by Crossrail (the claim was questionable in the case of Stansted, since accessing that airport via Crossrail would involve a change of trains at Liverpool Street). But (and there had to be a 'but'), the Department complained they lacked precise information about how much the project would cost. With contingency planning, the bill had risen to £10bn, but Darling did not trust this estimate. His scepticism about big projects had increased because the government had been badly stung by two recent projects.

First, the cost of the Jubilee Line Extension had caused an almighty furore as the eventual bill was £3.5bn, 85 per cent more

---

* I bumped into Darling coming off the Scottish sleeper soon after he left government and commented about how Scotland was investing in lots of major projects. He seemed appalled and commented that it was 'our money' that was being spent, a reference, I think, to the UK government supporting Scotland.

than the original estimate of £1.9bn. There were good reasons for the overspend: these included a deliberate underestimate of the projected budget at the outset, in order to get the scheme past the Treasury; the collapse of a tunnel during construction on the Heathrow Express in October 1994, which meant that work on the Jubilee Line Extension – which was being carried out using the same New Austrian Tunnelling Method* – had to be halted while investigations were carried out; and the particular difficulties associated with meeting a deadline of the last day of 1999.[†]

Secondly, another project – the upgrading of the West Coast Main Line – had gone even more haywire. The overspending on this scheme, intended to increase the capacity of the line linking Euston with Birmingham, Manchester, Liverpool and Glasgow, was off the scale, even by the standards of megaprojects generally. Originally estimated to cost £2bn, the projected spending on the scheme was now £9bn, and, during the period when the Crossrail Business Plan was being discussed the bill, seemed to be mounting daily. Estimates eventually reached £14.5bn. It was only a year later, when the SRA stepped in to rein back the huge overspend, that the final bill was brought down to £8.6bn – and this only after much scope reduction. These examples of massive overspending on projects in his department made the

---

* A form of spray concrete lining which is explained in greater detail in Chapter 10 (p. 212).

[†] Such niceties tend to be ignored by politicians who are unwilling to take the plunge on big-ticket projects, but it is interesting how these 'scandals' are quickly forgotten when the new piece of infrastructure becomes so heavily used that everyone wonders what the fuss was all about and the cost overruns are simply a few extra million on the national debt.

ever-cautious Darling even more wary of setting off another budgetary explosion.

Politicians with Darling's mindset have a powerful weapon up their sleeve to help them wage war against the adoption of high-budget projects. One such weapon is the very method by which the cost of these projects is assessed, which reduces their chances of obtaining government support. The gross figures of the cost announced at the outset fail to take into account the fact that a large proportion of the expenditure goes straight back into government coffers in the form of various taxes and reduced payments. In my book about the Treasury's public–private partnership for the London Underground, I quoted a memorandum sent to me by London Underground: 'Because the project started in the middle of a recession, it is credited with creating 50,000 jobs and the Treasury was able to claw back 47 per cent of the final cost in corporation tax, national insurance and income tax, and unemployment benefits that did not need to be paid out.'[14] Taking such payments into account – i.e. calculating expenditure on the basis of net rather than gross – would be a much more accurate way of assessing the cost of these schemes to the Exchequer, and would have the effect of making many seemingly unaffordable projects viable.

Another weapon available to the parsimonious politician is the Treasury's faithful best friend – delay. Darling's predictable decision after a few months of little action, apart from the inevitable round of public consultation, was to announce that the scheme needed to be reviewed. The man chosen for the job, Adrian Montague, was a solicitor and businessman who was well known in Treasury circles as he had headed the private finance initiative task force in the late 1990s, which had developed the Labour government's extensive and controversial PFI programme. It was not difficult, therefore, to discern Darling's intentions.

The Montague Review was a classic case of government double marking as it was essentially charged with justifying the project all over again. Ironically, as is often the case with such reviews, most of the work was carried out on Montague's behalf by the same team that had drawn up the *Business Case*. Montague's broad remit was to assess the 'full cost' of the *Business Case* proposals and, in particular, to judge whether they could be delivered on time and offered value for money. Crucially, he was to determine 'the proportion of funding required from non-Government sources'.[15] In other words, the promoters would have to seek money from sources other than the Treasury.

Montague reported exactly a year after the submission of the *Business Case* and broadly supported the scheme: 'Overall, the review has concluded that the *Business Case* for the Benchmark Scheme is Robust [*sic*] and could provide good value for money.' This statement was followed by the banal caveat: 'It is, however, critically dependent upon the extent to which the expected population and employment growth in London materialises.'[16]

Inevitably there were numerous provisos. Montague's report expressed doubts as to whether the twenty-four trains per hour operation could be achieved, in the light of the difficulties of interacting with the national rail network outside the tunnels. Rather oddly, since Montague had no expertise in rail operations, the report explored the feasibility of this level of service in considerable detail, suggesting – in a curious piece of nit-picking – that levels of twenty to twenty-three trains per hour rather than the proposed twenty-four would be possible.

In assessing the costs, the report found that some had been overestimated while others had been underestimated, and that broadly the two variations cancelled each another out. With the addition of substantial contingency – the optimum bias requirement had not yet been imposed – the overall cost of the scheme was reckoned to be around £10bn. Two measures to save costs in operating the railway had been introduced in the *Business Case*. The original plan that the trains would have guards on board had been scrapped and this would, the report said, lead to substantial savings in operating costs.[*]

Another change proposed by the *Business Case* was the scrapping of the idea of making the trains capable of taking electricity both from overhead wires and the third rail system which is used on the existing tracks out to Kingston and throughout the southeastern commuter network. The idea was to convert the tracks rather than buy trains designed to work with both systems, but it was soon apparent that the plan for the Kingston branch was toast. The Kingston branch had been the SRA's favourite option for Crossrail in the west, since it would relieve overcrowding at a third London station, Waterloo, but Montague's review, however, undermined the case for it: 'plans for the Richmond–Kingston branch are still relatively underdeveloped; coincidentally, this is also the aspect of the current route design that has attracted most public opposition.'[17] Jim Steer, who helped draw up the *London East-West Study* while deputy director of the SRA,

---

[*] Industrial action arising from the introduction of driver-only operation would bedevil services on Southern and other networks in 2016–18, so Crossrail's early decision looks to be a fortunate one. It is also one that cannot be reversed, as the trains have been designed for one-person operation.

confirms that there was considerable local opposition: 'It came mainly from people in Richmond who would lose a few feet of their back gardens and who were also concerned that they would lose District Line services which would be replaced by Crossrail.'[18] However, as Steer points out, the Kingston branch had by far the best benefit–cost ratio and was relatively simple to build. It would have needed a tunnel in the eastern part of the route, but thereafter it would have been able to share tracks with the North London line, passing through another short tunnel to connect with existing suburban lines under Richmond town centre and therefore its cost was relatively modest.

The estimated cost of £890m would have delivered benefits of £3,528m, giving it a benefit–cost ratio of almost 4–1, double the ratio of the benchmark scheme between Paddington and Liverpool Street. Despite the good business case, and despite the fact that Labour voters in the affected areas are few and far between,* Darling killed off the Kingston branch in his response to the Montague Review when, replying to a question from a fellow Labour MP, he told the House of Commons: 'He [Montague] recommending [sic] dropping – we are doing so – the line into Richmond, which will save about £1 billion.' This was, in fact, disingenuous since the benefits associated with the Kingston branch were so high and Montague had certainly not advised killing it off altogether. He had been decidedly lukewarm about some aspects of the idea, suggesting that interfaces with the existing railway should be limited, but only suggested that 'removal of the Richmond/Kingston branch... would make a

---

* Something with which the author, as a former Richmond Park by-election candidate for Labour, is all too familiar!

significant contribution here'.[19] Darling accepted that dropping the Kingston branch would reduce the overall benefit–cost ratio of Crossrail, but 'not so much' that it would make the whole scheme unviable. Darling's decision was plainly motivated by penny-pinching – a desire to knock a billion pounds off the overall cost, even though the benefits were far greater, another example of a politician ignoring the cost–benefit methodology because of his particular world view.

The alternative was to send trains down the Great Western line to Maidenhead (this would – much later – be extended to Reading). The cost of this would be far lower, since it required electrification of the line only from Heathrow Airport Junction, an expense that would be more than matched by extra revenue and the cheaper cost of running electric, rather than diesel, trains and, as already mentioned, thanks to the later decision to electrify the Great Western line, not a cost that would be borne by Crossrail. Nevertheless, the scrapping of the Kingston branch, as well as the Metropolitan link, has left something of a hole in the completed Crossrail scheme; instead of being used to connect with other areas, half the westbound trains will turn around at Paddington because they have nowhere to go. In hindsight, this was the worst and most short-sighted mistake made in the planning of the scheme and one that is bound to be reversed in some way by ultimately providing further destinations in the west, even though Kingston may well not be the beneficiary.

Indeed, critics of the scheme, such as Michael Schabas, argue that this is a wasted opportunity that makes insufficient use of the expensive tunnels. In 2004, in response to the publication of the Crossrail Business Case, Schabas and a group of railway managers produced a detailed proposal called Superlink. This offered an alternative vision of the services running through the cross-London tunnel – one that was much more akin to the Thameslink concept than the planned design for Crossrail. The Shenfield branch, they argued, was a waste of resources since it did little to improve services on the Greater Anglia line, apart from helping a few people commute directly into the West End. Schabas and his colleagues put forward a range of alternative destinations. In the east, services from the tunnel would serve Cambridge, Ipswich, Southend, Stansted and Pitsea (in Essex), while in the west they would go to Guildford, Basingstoke, Heathrow (with through trains to Reading) and Northampton, via Milton Keynes. Schabas and his team had succeeded in focusing on Crossrail's Achilles heel, namely the limited number of stations it serves and the difficulties of finding destinations in the west.

Superlink had no official status and the plan had not been properly worked through, but it attracted considerable publicity. Crossrail's backers, worried that it would lead to yet more delays, saw it as a threat and commissioned a couple of reports to debunk the idea. Schabas characterizes these as 'a smear campaign', but they were in fact researched and written by respected railway managers who identified the potential pitfalls of his plan. They pointed out that the large number of destinations on either side would make it impossible for trains to run at intervals of 2.5 minutes through the tunnel as they would be subject to delays on the tracks leading to it. Superlink's promoters argued that the RER in Paris and the S-Bahn in Munich are able to cope with such multiple destinations. However, Howard Smith, who formerly worked for TfL, is adamant that the situation there is very different: 'Munich does have five branches at one end and six at the other, and indeed does run thirty trains per hour through the main tunnel. However, their overcrowding is nothing like as severe as in London and they have Spanish-style platforms – i.e. where you get people entering on one side and exiting on the other, which is far quicker.' Smith visited Germany and France specifically to see how these systems function and was convinced that would not work in London: 'They also have stations with multiple platforms at the entrance to the tunnels which allows them to regulate trains into it, whereas we do not – indeed, Thameslink is going to have great difficulty pulling off this trick without any of these factors.'[20]*

Another criticism levelled at the Superlink scheme was its attempt to marry two concepts – regional and metro services – that were essentially incompatible. Gordon Pettit, author of one of the reports assessing Superlink, explained: 'In the UK operators must provide seats for passengers where their journey time is more than twenty minutes. There is no doubt that you get a better financial result by running longer distance trains through the core tunnel, but the trains you require do not have the capacity you need for people standing in the busy cross London section.'[21]

As mentioned before, regional services on which people travel for up to two hours require a higher standard of comfort and a greater number of seats than commuter services. They also require toilets and a refreshment bar, neither of which Crossrail's trains will have. Trains designed for a regional service have

---

* Smith, speaking in March 2018, was being prescient. At the time of writing, June 2018, the introduction of the new Thameslink timetable was causing chaos on the rail network, with large numbers of cancellations and long delays.

carriages with fewer doors, whereas those who planned Cross-rail's rolling stock were adamant that each carriage must have three, rather than two, doors on each side, so as to ensure short dwell times in stations.

The other objection to the Superlink scheme, with its focus on serving regional rather than suburban destinations, was that it did not address Crossrail's main purpose of relieving over-crowding on parts of the Underground and at mainline stations.

Both sides' arguments have their merits. Crossrail's initial operation plans, involving a number of trains turning round at Paddington, remain flawed and undoubtedly will be changed over time – even if this requires further investment. Indeed, some of Superlink's suggestions may well be part of that process. However, as the then mayor of London, Ken Livingstone, pointed out, Superlink came to the table too late – in 2004 – and he did not want anything to interfere with Crossrail's progress. Schabas accepts that 'Livingstone was right. Superlink broke cover too late. Policy makers were suffering project fatigue and although many privately agreed that Superlink did indeed seem to be a better scheme, most found disagreements between "experts" to be tiresome.'[22] Schabas admits, too, that he made two mistakes. First, he questioned the alignment of the tunnels, because he wished to create a route under the river but in fact his proposed route would not have served as many central London destinations as Crossrail's, and moving away from the safeguarded route would have resulted in further delays. His second error was to say – repeatedly – that the scheme as set out was 'unfundable' when, in fact, it was not.

By now, Crossrail had attracted the support of the prime minis-ter, Tony Blair. In an interview with the *Financial Times*, he was eager to show his business-friendly credentials, stating Crossrail 'was an investment [that] we can't afford not to make'.[23] Darling, who clearly had to be cajoled into supporting the scheme, reluctantly announced that a hybrid bill would be introduced 'at the earliest opportunity to take the powers necessary for Crossrail to be built'.[24] However, he stated that a 'major funding challenge' remained to be resolved before the project could go forward and that the onus would be on 'those who benefit from Crossrail' to 'contribute substantially to its delivery'.[25]

Indeed, the Treasury continued to have its doubts. On the day the Montague report was published, the *Daily Telegraph* had quoted a Treasury source as saying: 'We are a long way from any resolution. In particular we need to consider the project in the wider context of the funding pressures on the national rail network, and, as the CBI indicates, business is still in the early stages of considering any contribution it might make. These are not issues that can be quickly resolved, but Adrian Montague's report will help us make further progress.'[26] The next battle to be fought, therefore, was over the money.

6.

Seeing off the naysayers

As befits a Treasury-inspired initiative, the main concern in Montague's report was the money. The project could not, he argued, all be funded by taxpayers and he identified four potential sources of funding other than government grants. Two of these – European funding and a fares premium – were quickly dismissed as unrealistic. This left two options: obtaining a contribution from London, whether businesses or individuals, in the light of increases in property value arising from the proximity of the new railway; and, more directly, allowing Crossrail to profit from the development of land. Montague's view was that 'localities within one kilometre should be likely to benefit and might therefore be expected to contribute to the costs of Crossrail'.[1] According to Montague, Crossrail's property advisers reckoned that an analysis of previous schemes showed the increase in values would be in the order of 6 per cent, though with sizeable variations depending on the locality.

Transport for London had already carried out considerable work on a potential means of extracting value known as 'tax increment financing'. As Montague explained, 'tax increment financing allows development projects to be financed with incremental property tax revenues generated by the increased property values expected to result from the new development'.[2] It is a complex method. The amounts paid are dependent on the relative property value movement between those benefiting from the development and those who are not. This muddies the waters since there can, of course, be numerous other factors affecting price movements that have nothing to do with the development, which means it is a fairly hit and miss method. TfL estimated that, if introduced quickly, with the business rate revaluation of 2005, as much as £3.2bn could be raised in this way, provided the extra payments were maintained until 2039.

The alternative suggestion – a version of which was eventually adopted – was the introduction of a blanket supplement on the business rate in the area that would benefit from Crossrail. This too would be imposed until 2039, but again there were complexities over such issues as defining the boundary and determining which businesses – or types of businesses – should pay. The potential gains were, however, large. TfL estimated that around £2bn could be raised from businesses based just one kilometre around Crossrail stations (the double entrance to all the stations in the central area of course greatly increased the catchment area) if the extra tax were imposed between 2005 and 2039; and expanding the area could bring in up to £1bn extra.

The other potential source of income suggested by Montague was through property development, either above stations on land owned by Crossrail on sites nearby. Montague criticized the authors of the *Business Case* for not being sufficiently ambitious in this regard. He pointed out that while the *Business Case* included a discussion of the potential of such opportunities, it suggested that there would not be enough revenue to include the amount in its calculations. This was a mistake, he suggested, since 'there is a popular view that property development can be used to create sufficient value to fund (or make very significant contribution to) the costs of transport infrastructure'.[3]

Montague therefore considered in some detail whether property development could make a substantial contribution. He thought the prospects on development over stations were limited because any acquired land would legally have to be for construction rather than for development. Obtaining land from neighbouring landowners would be difficult as Crossrail would not have compulsory purpose powers and therefore any development opportunities would, he reckoned, require cooperation with these neighbours.

The uncertainties of the property market and its cyclical nature made it difficult to assess potential receipts, but Montague reckoned these station sites could generate profit of between £100m and £300m. This proved to be rather conservative as, in the event, the current estimate is that, thanks to booming property prices in London, development at a dozen sites in central London will raise £500m for Crossrail's core funding.

Had legislation in the UK been more favourable, however, this figure could have been much higher. The main obstacle preventing Crossrail benefiting from development of other sites along the route was that 'there is currently no statutory or fiscal mechanism by which development gains could be collected directly by Crossrail'.[4] This is a major lacuna in the current planning system – and one that will need to be changed by a future government.

Instead, Crossrail was for the most part dependent on contractual obligations entered into voluntarily by developers (clearly with some pressure being applied) or through agreements with the local authority. Under Section 106 of the Town and Country Planning Act 1990, developers pay for the creation of local facilities in return for obtaining planning permission. Montague's report was produced before the introduction in 2010 of the Community Infrastructure Levy, effectively a tax on planning which was more comprehensive than Section 106. Both Section 106 and the Community Infrastructure Levy ultimately contributed further significant amounts to Crossrail.

Montague barely gave any consideration to adopting a private finance initiative arrangement to fund Crossrail – despite the fact that PFI was currently all the rage. PFI was being used to finance hospitals, roads, schools, defence contracts and any other government investment projects and, indeed, had been part of his remit. The idea behind PFI was that the private sector could invest capital in state projects for a high rate of return, but with the proviso that it also took on some risk in the event of overspend or failure. Public bodies can borrow more cheaply than private companies because they have the certainty of state backing behind them, but the higher cost of borrowing for private organizations would, as far as supporters of PFI were concerned, be more than compensated for by the transfer of risk and the greater efficiency of private companies. The Blair government developed an unhealthy addiction to these schemes and had just negotiated one of the largest public–private partnerships (PPP, a slight variant of the PFI) for the maintenance and renewal of the London Underground – a deal which would collapse a quarter of the way through its thirty-year contract.*

Montague pointed out that a PFI deal would only be worthwhile if risks were genuinely transferred to the private sector and 'remain with the private sector and not revert to Government under pressure of failure'.[5] Recognizing that the cost of capital rose as the risk increased, he observed that, in the case of Crossrail, the 'sheer size and complexity of the infrastructure means that the private sector is poorly placed to assume construction and revenue risk'.[6] He concluded, therefore, that 'it would not represent value for money to utilise private finance'.[7] Crossrail would later attempt to use PFI for rolling stock procurement, but, as we shall see, this also was adjudged not to offer value for money and the idea was dropped.

This protracted search for new ways of funding transport infrastructure was a radical departure for a British transport project. Historically, with the odd exception such as the expansion of the Metropolitan railway lines into Metroland in the first half of the twentieth century, there had been very few attempts to link private profits made from transport developments to the public expenditure on building them. It is interesting to note that even in the USA, supposedly home to the most purely capitalist approach to economics, tax increment financing is widely used. Legislation to enable money to be raised in this way, which was first used in California in 1952, has been passed by all fifty US states. Nor are these methods particularly controversial, though

---

* As recounted in my (presciently titled) book on this disastrous contract, *Down the Tube* (2002).

inevitably there are sometimes rows over increased taxes. While no company likes paying extra tax, Montague stressed that these alternative methods of funding were supported by London business interests and their representative organizations. The UK is well behind the game on this issue and the situation here is in sharp contrast to that of many other countries, most notably Hong Kong, where the state organizations that develop transport schemes are also able to benefit hugely from associated development.

In tandem with the Montague Review, a massive consultation exercise took place involving major 'stakeholders' (a buzz word of the Blair government) such as local authorities, key environmental bodies like English Heritage and English Nature, and the general public. Winning the public over to Crossrail was seen as crucial by its promoters to avoid another debacle. Numerous meetings were held along the route and a mobile information centre was taken to various locations so that local people would have easy access to the information, which was provided in eleven languages, as well as in Braille, large-print and audio cassette versions.

After the publication of the Montague report, there was a second public awareness campaign followed by another tour of public information centres, exhibiting the more detailed designs that had by then been developed. Local businesses and residents were invited to information centres at fifty-five locations along the route over a period of several months in 2005. Crossrail claim to have attracted 15,000 visitors, a relatively impressive figure but only a tiny fraction of those affected by the construction of the railway. There was, too, the usual telephone 'help desk' which was staffed even at weekends and was contacted by a further 11,500 callers. Taking place in the Internet age, all this was backed up by websites and social media, the first time such a thing had happened in the UK for such a major project. The Crossrail website, www.crossrail.co.uk, contained a vast amount of information and presented all the relevant documents, including the massive *Environmental Statement* 'setting out the likely impacts on the environment arising from the project'; it attracted some three-quarters of a million visits between being created in 2001 and the start of the parliamentary debates five years later.

With the government generally supportive of Crossrail after the Montague Review, the consultation process found that the public, too, with some inevitable exceptions, was in favour. Consequently, in the expectation that viable sources of funding for the project would be found, work continued on the hybrid Crossrail bill. With the help of a further £100m from the government on top of the original grant of £145m, the bill was finally deposited in Parliament in February 2005. In April the transport minister, Tony McNulty, set out the scope of the bill. While he provided an overview of the size of the task and the challenges facing those who would be charged with delivering the scheme, he also made a plea to ensure that the bill would be carried over to the next Parliament after the coming general election. The minister stressed that Crossrail was a scheme of national, rather than simply local, importance. This was clearly an appeal for support from MPs in the regions, and was also evidence that lessons had been learned from past failures, which had been caused partly by the perception that the scheme only benefited London and the South East. It was important to sell the scope of the project – especially since it was yet another case, following the Jubilee Line Extension, of a huge amount of money being spent on a scheme based in the South East, where expenditure on transport was so much greater per capita than in the North. Crossrail was national, McNulty said, because 'London plays a major role in supporting regional economies and jobs, through commuting, product and service purchases, fiscal transfer and economic activity'.[8] That may well be so, but McNulty's words did little to reduce dissent from some northern MPs who felt that spending was already disproportionately slanted towards the South East, an issue that has gained more prominence in the intervening decade and threatens to derail plans for Crossrail 2.

McNulty went on to outline the amount of work that had been done in preparing the bill:

> When the Bill was deposited, it was supported by a regulatory impact assessment, a race equality impact assessment, a book of reference, an estimate of costs, an environmental statement and a non-technical summary of that statement, parliamentary plans and sections, a European Court of Human Rights statement, a housing statement and explanatory notes. Some 4,600 landowners' notices were served and approximately 400 street and footpath notices were put up. The environmental statement comprises some 3,700 pages and nine volumes and is supported by a further 14,000 pages of specialist technical reports. The book of reference contains more than 5,000 entries, which would need to be re-checked if the Bill had to be deposited again because the book of reference must be no more than 28 days old at the time of deposit.[9]

This vast array of material was the result, McNulty said, of four years' work carried out by Cross London Rail Links, which employed around 100 staff and 90 consultants. He then went on to describe, in a rather light-hearted way, the sheer scale of the consultation exercise: 'For those who are interested in such data, around 14 tonnes of material were distributed to 140 different locations to meet the requirements of Standing Orders, which include ensuring that documents are available for public inspection locally.' He omitted to say how many trees had been sacrificed to provide them.

The bill confirmed the route and, crucially, the alignment of the tunnels under central London. The tunnels, which would be twin-bore – meaning that the east- and westbound bores are separate from one another – would run through the whole central section and part of the southeastern alignment towards Abbey Wood.

In the west, the tunnel starts between Royal Oak and West-bourne Park, the first two stations out of Paddington on the Hammersmith & City Line. Therefore, by the time the line reaches Paddington, the tracks are well below ground level, but Paddington's Crossrail station has been designed with a transparent roof that will allow daylight down onto them. The next station is at Bond Street* below the Jubilee Line with which it connects and, as with all the central stations, there will be entrances at both ends – at the east end it will be in Hanover

---

* Funnily enough, Bond Street does not exist since there is a New Bond Street and an Old Bond Street but no plain, simple Bond Street.

Square, a stone's throw from Oxford Circus where Crossrail will not have a station. As a result there will be no direct connection with the Victoria Line, which will surprise and inconvenience many passengers. London Transport decided early in the process that Oxford Circus was already so overcrowded that a Crossrail link would make matters worse, but this is, on the face of it, rather counter-intuitive thinking as it implies that a facility should not be provided if it is expected to be overused. The problem, London Transport believed, was that expanding Oxford Street station would have resulted in the demolition of one of the iconic buildings on the corner of Oxford and Regent streets. The space available in the area is undoubtedly limited, but as Michael Schabas puts it: 'A quarter of a century later capacity on the Victoria Line is being increased with new trains and signalling, and Victoria Underground station is being upgraded at great expense with a northern ticket hall and a deep level concourse. Perhaps something similar might have worked at Oxford Circus.'[10]

Tottenham Court Road is so close to Bond Street that trains will barely get up any speed between the two, but it was felt necessary to ensure both ends of Oxford Street were served by Crossrail. Again, there will be two entrances: one in Dean Street in deepest Soho as well as one near the existing Tube station, which has been completely rebuilt and expanded. There will be subterranean connections with the Northern Line as well as the Central. The Cross London Rail Links team considered the idea of having a station at Holborn, but apparently the business case was poor and the idea was not progressed. To reach the next stop, Farringdon, the route briefly departs from a parallel alignment with the Central Line and connects there with the

Circle and Metropolitan lines. Even more significantly, the national rail station at Farringdon is also a stop on Thameslink, which has been greatly expanded as a north–south cross-London link with many new destinations such as Cambridge, Rainham and Horsham. The station will quickly become a key hub for the whole London suburban network and one of the few in the national rail system which will have numerous destinations to all four compass points. Moreover, Farringdon will run direct trains to Heathrow, Luton and Gatwick airports, providing a convenient interchange between them. It is not inconceivable that Stansted may one day be added to that list. A massive concourse, several times bigger than the original Farringdon Underground station, has been built to cope with the thousands of people expected to interchange there. As with the other central London Crossrail stations, there will be a second entrance, right next to Barbican station which lies at the eastern end of the extensive Smithfield Market – a demonstration of the sheer length of the trains that will run on Crossrail.

Farringdon was until recently a rather dismal and neglected little station building whose frontage still bears its former name of Farringdon & High Holborn station, an inaccurate designation since it is nowhere near High Holborn, but it is now set to become one of London's busiest interchanges. There is one long-dead Victorian who may even be stirred from his grave by this amazing change in fortune for Farringdon. He is Charles Pearson, the father of the London Underground, whose efforts ensured the construction of the world's first Underground line between Paddington and Farringdon, but whose other project, to provide London with a huge central station at Farringdon serving all points of the compass, never materialized because of opposition from the City authorities. A century and a half later, his dream will be brought to fruition.

After Farringdon, Crossrail then follows a route directly under the Circle Line, which was the only possible alignment to avoid the piles under the huge post-war Barbican Centre. This means that Liverpool Street Crossrail station's western entrance will be at Moorgate, connecting with the Northern, Metropolitan and Circle lines and with commuter services on the Great Northern Line to Finsbury Park and various suburbs and outlying towns. To quote Michael Schabas again: 'In a sense, London's newest Underground line is being built in the "shadow" of the oldest.'[11]

Once the route of the tunnel beneath central London was determined, discussion within the Crossrail team focused on what services should link into it from either end. Don Heath was in charge of the alignment outside the tunnels, which required considerable difficult negotiation between him, representing British Rail, and London Transport and London Underground. London Transport saw the Crossrail concept as a metro line through London, while British Rail wanted a regional rail service stretching further out into the Home Counties and possibly even beyond. On the eastern side there was overcrowding at Liverpool Street station that needed to be relieved, while in the west Paddington was less crowded but still a potential bottleneck. Reducing the number of people interchanging at both of the two mainline stations covered by Crossrail was the core aim of the project. The obvious route for Crossrail after it emerged from the eastern end of the central London tunnel was to take over the existing services to Shenfield, an affluent suburb of Brentwood which has long been the terminus for the trains from Liverpool Street to outer east London and Essex. The trains on this route had become severely overcrowded because of Big Bang, experiencing a growth of 40 per cent in the decade running up to 1991 as the deregulation of the City had created jobs for baristas, brickies and clerks as well as bankers. In Michael Schabas's words: 'Essex had become known somewhat disparagingly as the "county of clerks" most of whom walked out of Liverpool Street station every morning and into a job in the City.'[12] Many also transferred at Liverpool Street onto the Central Line to get to the West End and it was these people whose journeys to work would be greatly improved by Crossrail. The commuter lines out of Liverpool Street were perceived by Crossrail's planners as the most overcrowded in London, and their radical overhaul constituted the most obvious reason for building the new line.

In the west, it was more difficult for the planners to work out what to do with the Crossrail trains. The link with the Metropolitan Line from a junction under Hyde Park was soon dropped as it was deemed too expensive and would have halved the number of trains going to Paddington. The planners realized it would be easier to link with the Metropolitan on the surface at Old Oak Common in Acton, where the West Coast Main Line and the Great Western are within touching distance of each other – though there are a few buildings in the way. However, the need for such a link was unclear, since the Metropolitan already provided regular services deep into the City via Baker Street, while those travelling to Canary Wharf would be able to access the Jubilee Line once the extension was completed. Despite this logistical flaw, a link at Old Oak Common, with trains going onto the Metropolitan Line and the Chiltern tracks out to Aylesbury, Amersham and Chesham was retained in the preamble to the Crossrail Bill setting out its purpose; it therefore constituted one of the key components of the scheme. The plan was that all the services from those Buckinghamshire destinations would in future be served by Crossrail. This would relieve pressure on Baker Street but would make some journeys much less convenient for commuters. There were also huge technical difficulties to overcome, as David Smith, one of the planners for the Crossrail scheme, explained: 'the Metropolitan line alignment was very narrow, which would have made it difficult to put up wires, and the track bed was not at all solid, not like Brunel's out to the West or the Great Eastern to Shenfield. It would have required a lot of work and hundreds of thousands of tons of ballast to shore it up. Even the rails themselves were not in a good state.'[13]

The link with the Metropolitan was in effect an attempt to find a purpose in the west for a scheme whose main focus was relieving congestion in the east. As already noted, Paddington has never been a big commuter destination like Waterloo or Victoria, let alone Liverpool Street; historically, there have always been many more local jobs in west London which in the past mostly did not require specialist skills and therefore damped down local demand for commuting. While the more recent emergence of high-tech firms in the Thames Valley has created some demand for travel, much of it has been on the relatively empty trains travelling against the peak flows. Another honeypot for jobs in this area is Heathrow, whose relentless post-war expansion has created considerable employment both around the airport as well as within it; oddly, however, at this stage the planners did not properly work up the idea of a Heathrow service for Crossrail. While the bill made provision for a link with the tunnels for the Heathrow Express service, which were then being planned, according to Schabas the Cross London Rail Links team 'had not got round to doing a deal with BAA plc, the privatised airport owner, which was concerned that Crossrail trains, with lower fares, would undermine the profitability of their Heathrow Express trains'.[14] This was another issue that had not gone away by the time the Crossrail scheme was revived.

Working out viable routes to the west therefore continued to be a headache for the planners. Existing commuters already had lots of options. Those arriving at Paddington could easily access the City using the Circle Line, while passengers bound for the West End had the benefit of the relatively lightly used Bakerloo into Oxford Circus. Admittedly, people heading for Docklands had rather more difficulty, requiring at least one other change of train, but the planners found it difficult to justify the massive increase in rail capacity that would be delivered by Crossrail in west London. None of the options for routing the trains beyond Paddington were obviously superior to the alternatives. Even as the scheme is nearing completion at the time of writing, there is still uncertainty about precisely what destinations will ultimately be served by Crossrail in the west. Initially, half of the twenty-four trains per hour at peak times in the westbound direction will terminate at Paddington – unquestionably a waste of resources and a failure of imagination in developing new destinations that would link parts of outer London or out-lying towns with rapid rail services to the centre of London and Docklands.

There was never any doubt that Crossrail trains would be powered by electricity since the idea of running smoke-belching diesels through the central tunnels was unthinkable. At that stage there were no plans to electrify the Great Western and consequently Crossrail would have to pay for the whole electrification of the railway between Airport Junction, where the electric Heathrow Express trains turn off the mainline for the airport, and Reading. Therefore, cutting back services to Maidenhead was felt to be sensible as it would have obviated the need to electrify the fifteen miles to Reading, saving a considerable amount of money. Once electrification of the whole Great Western to Bristol and beyond was announced by the transport secretary Lord Adonis in 2009, Crossrail could effectively get a free ride all the way to Reading and consequently it was announced that services would terminate there, rather than at the much smaller Maidenhead station.*

As previously decided, the tunnels would start at Royal Oak, just west of Paddington, pass beneath Hyde Park, the West End, Holborn, Clerkenwell, Shoreditch and reach Stepney. Here, in a change to previous plans, rather than coming to the surface, at a point beneath Stepney Green the route would fork – with one set of tunnels soon emerging to the surface at Pudding Mill Lane, just west of Stratford, and the other set of tunnels heading southeastwards, emerging near Victoria Dock Road in the Royal Docks. After a short distance on the surface, including the stop at Custom House Station, the route enters the Connaught Tunnel, an old tunnel previously used by the North London line, which was to be refurbished for use by Crossrail. After another short distance on the surface, the line would go under the Thames in

---

* The concept of bi-mode trains, capable of being powered either by electricity or by an on-board diesel engine had not yet been considered although they are now being introduced on the Great Western.

twin-bore tunnels, emerging just east of Plumstead station where it would run on the surface on new tracks adjacent to an existing railway alignment to Abbey Wood. In total, 46 km (28.5 miles) of tunnel were envisaged, though this was eventually reduced slightly to 42 km (26 miles), 21 km in each direction.

The routes at either end were, in fact, still subject to some change. At this stage, it seemed only Heathrow Terminals 1, 2 and 3 (which is one station) and 5 would be served, but later it was decided to add a service to Terminal 4 on the other side of the airport. In the east, the small town of Shenfield remained the terminus (as it has sufficient platform capacity); the other branch was confirmed as ending at Abbey Wood, a name that rather belies the dismal Thamesmead estate it serves. The idea of linking Crossrail with the Eurostar station at Ebbsfleet had been killed off: it was deemed that the likely traffic from the international railway would not justify the cost of a connection that would be technically more difficult to create than appears from the short distance shown on the map. The rather romantic vision, popular among some of Crossrail's early supporters, that it would serve international traffic, was therefore abandoned once and for all, though the possibility of extending the line remains if it is ever considered economically viable. Another more plausible option, of continuing services for a shorter distance through to Dartford, was also rejected on grounds of cost and convenience.

The tunnels were to be constructed at about the same depth as London Underground's Central Line, which is around 20 to 25 metres below street level through the West End and the City, though the line dips as low as 30 to 35 metres at a few points in its middle section. In the east, however, ground conditions mean that the tunnels are 40 metres below the surface between Liverpool Street and Pudding Mill Lane, and as much as 50 metres between the Isle of Dogs and the Royal Docks.

The bill also confirmed there would be Crossrail stations at Paddington, Bond Street, Tottenham Court Road, Farringdon, Liverpool Street, Whitechapel and the Isle of Dogs (or Canary Wharf, as it became known). The station platforms would be designed to accommodate the proposed ten-car Crossrail 20-metre carriages (in fact this became nine 23-metre carriages following Bombardier's successful bid for the rolling stock contract). However, the tunnels were to be constructed to allow for a future upgrade of platforms to 245 metres, for the operation of twelve-car trains. In addition to the tunnels, ventilation shafts no more than one kilometre apart would be required, though some could be incorporated into the station buildings.

The Crossrail team made a stab at potential passenger numbers. Rather optimistically predicting that the line would be in full use by 2016, the estimate was that 160,000 passengers would travel on the service during the three-hour morning peak period between 7 and 10 a.m. The most heavily used section would be between Whitechapel and Liverpool Street, with 55,000 passengers; Liverpool Street and Farringdon, with 49,000 passengers; and Paddington and Bond Street, with 35,000 passengers, confirming that usage would be far heavier in the east than the west.

The estimated cost was now £11.3bn, which included both contingency and the 'funding gap', a euphemism for calculating the subsidy required from the government. It was estimated that 1,360 jobs would be created in order to operate and maintain the new line, and that around 1,000 of these would be new jobs, with the others taken up by employees transferring from other rail organizations. During construction, however, the numbers employed would be far greater: the estimate was that it would be as many as 15,000 at the peak.

The nine-volume *Environmental Statement*, which was backed by a staggering 14,000 pages of technical assessments, considered in intricate detail both the negative and positive impacts of the project. It listed ten different sorts of impact, ranging from temporary noise from worksites to the effect of the new line on bus services. As well as addressing such obvious matters as air quality and contaminated land, the *Statement*, also considered transport issues, notably the impact of Crossrail on other major projects in the London area, such as Thameslink 2000 and the East London Line Extension. London had by then been selected to host the 2012 Olympic Games, and though it was known that Crossrail would not be finished in time the interaction between these two enormous projects had to be taken into account. The *Statement* was designed both to inform the MPs charged with examining the bill and also to provide evidence to objectors to the scheme who would be able to challenge aspects of the legislation in the committee hearings in both the Commons and Lords.

The *Environmental Statement* reiterated the aims of the project with a few refinements. Crossrail was intended to 'sustain employment and population growth in London; overcome constraints on the ability of Central London to continue to compete as a world-class financial centre; improve accessibility by public transport and promote sustainable development; and support regeneration, particularly in East London and the Thames Gateway'.[15]

To get a flavour of the detail which the consultants commissioned to write these documents were asked to produce, take the consideration of the 'route window' of Burnham, a relatively little-used station on the Great Western line where Crossrail trains will stop. Over half a dozen pages in volume three of the *Statement* detail the potential impact of Crossrail on Burnham station and its surrounding area.

We learn that Burnham station will need to have overhead line equipment, which is hardly surprising, but, more importantly, that the platforms will need to be extended by about 26 metres to accommodate the trains, which are longer than those currently using the station. Then the detail becomes rather more forensic: 'Platform extensions will take about four months to complete. Construction plant required at the worksite will include a crawler, excavators, mobile cranes, lighting rigs, compressors and generators.'[16] (A crawler is a kind of bulldozer.) Then there follows a description of precisely where and how the changes would be made: 'The works will be undertaken from one site at the western end of Sandringham Court, adjacent to the railway, and a second northeast of the station bounded by Burnham Lane to the east and Sandringham Court to the north.'[17] The *Statement* goes on to explain that there would be between two to four lorries serving each worksite and they would enter through Sandringham Court.

Next comes an examination of the exact location of Burnham station, including a statement that the area 'has potential for archaeological remains', but also that 'its more recent history of railway use is likely to have left a legacy of contamination'[18] (presumably of ash and unburned coal). There is 'a local nature reserve located northeast of the station but at some distance from the works', suggesting that it is unlikely to be much disturbed, especially as 'baseline noise levels are relatively high' given considerable local traffic levels and its proximity to Heathrow.

The next section looks at the immediate effect of the works on the area surrounding the station, which would 'require vegetation to be removed from a section of the embankment adjoining Sandringham Court. This vegetation is prominent in views... from nearby properties and its loss, together with the views of the construction works that will be exposed will have a significant impact on the visual amenity of about 25 properties in Sandringham Court.'[19] But happily, 'this vegetation will be reinstated on completion of works.' 'Vegetation' would also be lost from the lineside, including some woodland, but this, too, would be only temporary.

Finally, the *Statement* turned to the local wildlife: 'Surveys have identified habitat suitable for dormice at this location and further presence/absence surveys will be carried out prior to construction.'[20] A pond near the works was also to be surveyed to determine whether great crested newts were present. If either dormice or great crested newts were found in surveys, the *Statement* referred readers to the Appendix B1 – in a completely different volume – which set out at length the mitigation measures to be undertaken. And they are fantastically detailed. The best option is removal with release in their breeding pond, if its location is known, and, if not, at a suitable alternative site. However, even if the surveyors are not sure newts are present, but are concerned that they might be, the whole pond has to be cleared of any vegetation above 15 cm high and, quite literally, no stone left unturned in the search for these creatures. If any are found, work has to stop immediately and a licence sought from the Department of Environment, Food and Rural Affairs.

Newts can be the bane of project managers' lives. They joke that while newts are classified as a protected species – which suggests they are endangered – they turn up with unfailing regularity on railway schemes. Finding them can result in significant delays and, at times, huge costs.*

All this, remember, was for just one small section of the line. Multiply this lengthy and detailed process for the dozens of 'route windows' analysed in the *Environmental Statement* and one can begin to understand the bureaucratic process necessary to get a scheme such as Crossrail over the legislative and planning hurdles, and it is all too clear why the documents involved are so lengthy.

Ultimately, Crossrail seems to have got off lightly in terms of coping with inconvenient wildlife, since most of its new lines are in deep tunnels, well away from sensitive habitats. However, wild orchids, reptiles, newts and giant eels all had to be 're-homed' from around Stockley Junction, which is where trains turn off to Heathrow in London's far western suburbs, 'slow worms were moved from the Old Oak Common worksite' and 'fish rescues were carried out from the docks at Canary Wharf and the Royal Docks at Connaught Tunnel'.[21]

Another document the Crossrail team had to produce was a *Race Equality Impact Statement*. There were particular cultural sensitivities attached to this. Tower Hamlets, with its large Bangladeshi community, had played a significant role in opposing the passage of the first Crossrail Bill. Their concerns focused

---

* The project managers point out that while great crested newts are endangered in many parts of Europe, they are relatively common in the UK – hence the regularity with which they are found.

principally on the disruption that would be caused by a project from which they would derive no apparent benefit, since the initial Crossrail plan did not include a station in inner east London. This problem was partly remedied by the belated but welcome addition of Whitechapel to the route, which brought benefits not just to those living in the borough of Tower Hamlets but also to people interchanging between Crossrail and the Overground as otherwise there would have been no central direct link between the two, as well as with the District Line. In truth, the authors of the *Race Equality Impact Statement* struggled to find how particular ethnic groups were affected in ways that were any different from the wider community. The *Statement* highlighted the fact that Tower Hamlets was home to the largest concentration of minorities on the route, and mentioned that when an information centre was opened in Spitalfields, considerable efforts had been made to attract Bengali residents with fliers and banners. The *Statement* concluded that 'Crossrail will bring significant benefits to local populations and neighbourhoods through increased mobility for access to employment, health, education, cultural and leisure facilities. These benefits will extend across all the populations resident in the areas affected.'[22] However, most of the rest of the *Race Equality Impact Statement* outlined problems – and benefits – affecting all communities who lived in proximity to the line.

This brief summary of the preparation for the passage of the bill and the provision of information for objectors, demonstrates the sheer complexity and cost of putting a major project such as Crossrail through the modern planning process. When those opposed to a scheme ask why it is so costly to prepare a project, the answer can be found in this key part of the democratic process. It is undeniable that elements of the process could be carried out faster, more efficiently and with less of an obsession for detail – certainly the associated documents are all too often padded out with needless guff and banality. But the frequent demands to 'cut red tape' would result in a process that is less open, transparent and accountable.

The promoters are conscious, too, that Crossrail is likely to be part of London's landscape for the next century and quite possibly a good deal longer. After all, more than 150 years after the opening of the first section of the Metropolitan Railway, the very same bricks that the Victorians laid in the 1860s still line the tunnels.

The bill sat untouched in Parliament for a few months before the start of the key part of the parliamentary process, the House of Commons Committee stage, which involved sixty-three sittings lasting from January to October 2006. As with the failed private bill of 1994, a hybrid bill in effect grants planning permission to the promoters and consequently requires detailed and lengthy scrutiny. As before, the MPs selected to sit on the committee came from outside London so that they had no direct interest in it, and they were, like their predecessors, quite possibly selected because of past misdemeanours.*

---

* See Chapter 4, p. 74. The Chairman of the Commons Committee which considered the second Crossrail Bill was Alan Meale (Labour, Mansfield). The other committee members were: Brian Binley (Conservative, Northampton South), Katy Clark (Labour, North Ayrshire and Arran), Philip Hollobone (Kettering, Conservative), Kelvin Hopkins (Labour, Luton North), Siân C. James (Labour, Swansea East), Ian Liddell-Grainger (Conservative, Bridgwater), John Pugh (Liberal Democrats, Southport), Linda Riordan (Labour, Halifax) and Sir Peter Soulsby (Labour, Leicester South).

Crossrail tunnel under the Thames.

Before and after: the long-abandoned Connaught Tunnel in east London, built in 1878, has been refurbished in one of the most challenging undertakings of the project.

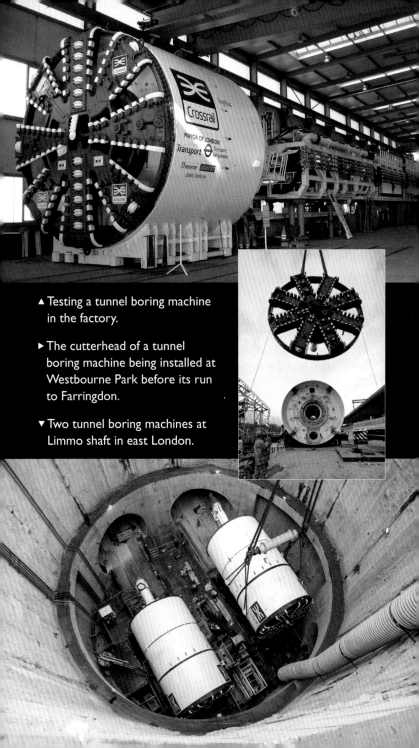

▲ Testing a tunnel boring machine in the factory.

▶ The cutterhead of a tunnel boring machine being installed at Westbourne Park before its run to Farringdon.

▼ Two tunnel boring machines at Limmo shaft in east London.

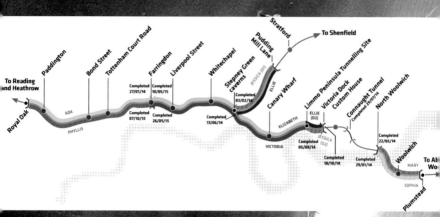

Map of the runs of each tunnel boring machine.

Tunnel works at Whitechapel.

Tunnel boring machine breakthrough at Whitechapel.

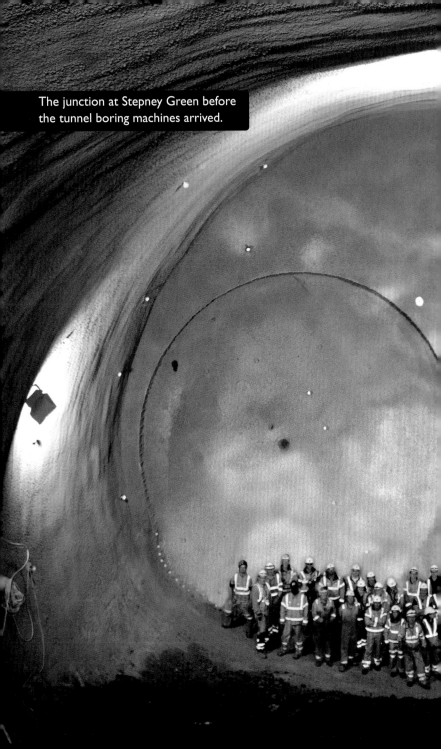

The junction at Stepney Green before the tunnel boring machines arrived.

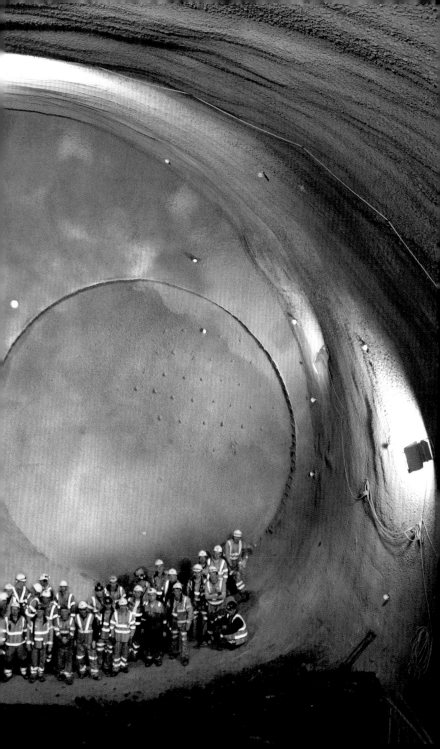

The huge concreting train ready to enter the tunnel.

The 250m long new passenger tunnel linking Paddington tube station with Crossrail.

Despite this, and again like their counterparts, they took matters seriously and made some important improvements to the bill. The fact that it was a hybrid bill did, however, impose constraints on the committee. In particular, the committee was not allowed to consider the principle of the bill. Furthermore, the government issued an instruction early in the process requiring that any environmental matters be referred to the full House of Commons, since they were deemed to be part of the principle of the bill. This so angered the committee members that they issued a recommendation at the end of the process to the effect that future hybrid bill committees should decide for themselves what constitutes the principle of the bill. However, because this was a hybrid rather than a private bill, there was no question of its rejection as the promoters did not have to prove the principle of the bill set out in the preamble. Nevertheless, petitioners who hoped to push through amendments to the bill or changes in the design or the way that the work was carried out raised an incredibly wide range of issues during committee stage. A total of 457 petitions were sent to the committee and its members considered that 205 of these were worthy of debate. Crossrail made considerable efforts to speed up the process: including, notably, going to the expense of installing a system of screens to enable the MPs, lawyers, petitioners and witnesses to reach the correct point in the documentation rapidly. (This was in order to avoid the constant reference to huge bundles of documents that is such a time-wasting and forest-destroying feature of court hearings.)

Issues put forward by petitioners ranged from matters funda-mental to the scheme to minor details relating to particular amenities. They were raised by large corporations and business

interests as well as, at the other end of the scale, residents' asso-
ciations and individual householders. For example, the first
significant point raised by petitioners was their concern about
the number of ticket gates for passengers using Liverpool Street
station. The two petitioners, the **City of London** and a big local
landowner, **British Land**, argued that not enough ticket gates
were being installed in light of the huge numbers of people
expected to arrive in the morning peak. Their presentations
soon plunged into detailed suggestions about the possible
location of Ticket Hall C, Tunnel J and the dozen gates. Much
of the committee's first session in January 2006 examined this
issue in detail.

Not surprisingly, therefore, the whole process required some
sixty sessions, spread over nine months and involving 17,500
paragraphs of evidence, to look at all the objections raised by
petitioners until the last day of petitions was reached. At that
final session in October 2006, the **Fairfield Conservation Area
Residents Association** in Bow, together with several other local
objectors, expressed their concern about the disruption caused
by the need to relocate a sewer in order to enable trains to access
Stratford station. In particular, they were worried about the
impact on Grove Hall Park, a local green space. After a lengthy
discussion, the matter was resolved by the chairman suggesting
that there be further meetings between the petitioners and the
relevant Crossrail team. While the committee was concerned to
ensure that Crossrail mitigated any damage, it also wanted to
see the promoters put forward positive proposals to improve
the park.

And, finally, there was a **Mr Anthony Chambers** whose house
was 13 metres above the tunnel near its eastern end in Pudding

Mill Lane. He complained about possible blight on his home and, specifically, the fact that he would not be able to install a heat exchange mechanism that would save on his heating bills because it would require a bore hole 60 to 70 metres deep that would clearly interfere with the tunnels. The blight, he reckoned, would knock £10,000 off the value of his home and the heat exchanger might have saved him as much as £12,000 over the next fifty years. Crossrail, he said, had offered him a measly £50.

The committee suggested that he either take 'his own legal advice' in respect of the issue of blight, or 'get in touch with the promoter' but the Crossrail QC did stress that, if he could prove that the value of the house had been affected, compensation would be paid. There was a suggestion that just possibly he may never have thought about using a heat exchange mechanism until it was suggested to him in order to prove detriment.

The couple of hundred petitions considered between these two hearings encompassed an extraordinarily wide range of issues. **G. E. Pensions Ltd**, which owned a property near one of the station sites in Oxford Street, wanted to ensure they would be able to have an over-site agreement to ensure the company benefited from any development (the committee dismissed its petition); the **Great Western Allotment Association**, whose members in Noel Road, West Acton, produce everything from 'potatoes to honey', were concerned about their proposed temporary relocation (they were promised that the new site would be tilled in advance of their move); the **London Borough of Havering** was concerned that the station at Romford did not provide sufficient facilities for mobility-impaired people (Crossrail promised it would); the **National Council of the Cycling Touring Club** sought to ensure that the interests of cyclists were taken into account by Crossrail (the MPs said that the design of the trains was up to the franchisee and not anything they could affect); members of the **Southend Arterial Road Action Group** were concerned that their houses would be affected by extensions to a work site at Gidea Park (further negotiations would be held); and three residents of **Bow** were worried about noise from the trains below their houses (they were promised that slab track would be installed, which causes less vibration); and so on.

Every petitioner's concern was addressed with great diligence even where it seemed the case was tenuous. So many commitments were made to petitioners to allay their concerns that a register of undertakings and assurances was created, enabling petitioners to check whether the guarantees had been adhered to once construction started. Interestingly, some of these agreements made between the Department for Transport and petitioners were confidential and in these cases only a summary of the arrangement was entered into the register. The House of Commons select committee, in its subsequent assessment of how the process had worked, expressed the view that it was hard for petitioners to check which commitments applied to them and to what extent their complaints had been engaged with. It is inevitable that some people will be left dissatisfied by such a process, but the fact that no judicial reviews were sought by any of the petitioners, which included major companies as well as individual householders and small-business people, is testament to the thoroughness of the process. China – where they send in bulldozers and boot people off their land without so much as a by your leave – this is not.

The biggest issue raised at this stage, and the most significant change arising from the committee's deliberations, was the addition of a station at Woolwich. This had been the focus of a long campaign led by the local MP Nick Raynsford. The decision was made following the guarantee of part-funding of the new station by Berkeley Homes, which planned to build a considerable number of houses in the area. The financial arrangement became part of the overall funding for the scheme which will be discussed in the next chapter. The seemingly simple issue of whether or not to have a station at Woolwich is an excellent example of how decisions made early in a project's history can later have a major impact.

Originally, when the SRA was involved in developing the scheme, it had been assumed that freight trains would use the central tunnels to get from one side of London to the other. While this might initially have seemed like a good idea – the few rail links between the two sides of London are already pretty full, and there is public enthusiasm for the idea of getting 'more freight on rail' – the practical obstacles are difficult to overcome. For a start, the Crossrail tunnels have a signalling system (about which much more later) called CBTC (Communications-Based Train Control), which is used nowhere else on the rail network and therefore the locomotives would have to be fitted with extra hardware and software. In addition, the locomotives would have to be electric, since it would not be possible to route diesels emitting smoke and dangerous fumes through the tunnel. A further problem is that there are no spare train paths available during the peak and few during the day generally. The freight trains would have to operate at night, greatly constraining the potential market and disrupting the tunnel's maintenance programme. On top of all this, freight trains are slow and heavy, and require gentler gradients than the lighter passenger trains.

That's where Woolwich comes in. Since the station is relatively near the end of the tunnel under the Thames, the original alignment of Crossrail, which was designed to accommodate freight trains, would have meant Woolwich station being deep underground, making it very expensive to build. Instead, during the parliamentary process it was accepted that freight trains would not be using the central tunnel section and therefore the alignment could be redesigned to allow a steeper incline between the southeastern end of the Thames tunnel and the proposed Woolwich station. That, in turn, made the whole proposition much cheaper as the station could now be built just below the surface through a process of 'mining' – in other words excavating from the top, rather than boring from below. Consequently, as a result of the lower building costs, the argument in favour of a new station at Woolwich became a compelling one. Woolwich is a bus hub, for services covering a swathe of southeast London, and consequently many local residents will have easy access to the new railway. The amendment to the bill to allow for the new station was therefore passed in May 2007.

Several other major changes were made to the bill during the committee process. There will be an additional 'ticket hall'* at Broadgate, one of the Liverpool Street entrances, because of

---

* A very outdated expression used by Crossrail, since barely anyone uses tickets now. It means a passenger concourse and exit.

the concerns mentioned above from the City and British Land about overcrowding caused by people using the station at rush hour. There was, too, the commitment that London Underground would build another entrance at Bond Street station to cater for the anticipated large numbers of passengers going to Oxford Street.

An issue that prompted controversy was Crossrail's initial decision not to install step-free access at all the surface stations served by the new railway. While it had been decided at the beginning to ensure all the stations in the centre were step-free, the policy was only to upgrade the other stations when passenger numbers increased sufficiently to justify the investment. This bit of penny-pinching came under fierce attack from disability groups, resulting in widespread media criticism. TfL quickly changed the policy, promising that all stations would be step-free by the time the line opened.

The committee's deliberations in response to petitions also led to a change in the location of the main train depot. This will now be at Old Oak Common, rather than Romford. The west London site offers more room and will be operationally more efficient, since it will reduce the number of empty stock movements.

The petitioning process was then repeated in the Lords, where a similar committee considered those petitions that had not been resolved in the Commons. Although petitioners are not supposed to put matters in a Lords petition that they could have pursued in the Commons, they can do so if the promoters agree to provide a 'Second House Undertaking', designed to allow more time to negotiate between the end of the Commons Committee and their appearance in the Lords. There were 113 petitions against the bill at the Lords stage and 45 of the petitioners made an appearance.

They were rather a mixed bag, with some of them plainly trying it on in order to obtain extra compensation – the rules were very distinctly laid out by Crossrail in advance and, given the large number of people affected, there were relatively few complaints in relation to compensation – or to try to get terms amended. For example, **Canary Wharf** made an unsuccessful attempt to restrict the amount of time that Crossrail was allowed to negotiate the compulsory purchase of a piece of land. In general, the Lords seemed to take a hard line with petitioners who had not been satisfied with the resolution of their complaint in the Commons and therefore made few changes. **Smithfield Market Traders**, whose site is bounded by the two entrances to Farringdon's Crossrail station, were worried about their ability to keep trading given the level of dust caused by station works on the surface but the Lords found that the commitments made by the promoters following their original petition were satisfactory. **Iver Parish Council** in Buckinghamshire, **the Ramblers' Association** and **the Open Spaces Society** petitioned for a replacement for Dog Kennel Bridge near Iver, which had to be demolished, but the Lords turned them down, saying there was no evidence that the existing bridge was being well used but they did recommend, if it were possible, for a footpath to be created to establish a way for locals to cross the railway. A **James Middleton** argued that the Crossrail scheme took the wrong route; this was viewed as a challenge to the principle of the project and therefore his idea was rejected – it was, indeed, strange he had got as far as appearing before the Lords committee. And so on.

The Lords, having to deal with far fewer petitions, only sat for twenty-nine sessions between January and May 2008. And that, apart from the later stages of the bill at which the various amendments were accepted, was that. The Crossrail Bill became law on 22 July 2008. A couple of days later Crossrail Limited – which had until then been formally known as Cross London Rail Links, and which (since the abolition of the Strategic Rail Authority in 2006) was now run jointly by Transport for London and the Department for Transport – was appointed to build the scheme, with an expected target opening date of 2017.

Simon Bennett, who was involved throughout the parliamentary procedure and is Crossrail's Head of Learning Legacy, says that 'the scheme presented to the next stage of the process [the Lords select committee] had been significantly improved'.[23] He notes that 'some of the changes to the project which resulted from negotiations with petitioners... could have been identified earlier and addressed before the bill was deposited.'[24]

Crossrail had, at last, been given the go-ahead. But where would the money come from to build it? Tony McNulty, in his speech just before the 2005 election, had referred to 'the little matter of £10bn' and accepted that there was still no clarity over the funding, although he stressed the government was keen to push on with the legislation. However, throughout the period of the committee stage there remained uncertainty about the precise funding of the scheme. In the short term, however, more problematically, there was still a lot of politics to come despite the parliamentary hurdle having been successfully jumped. The fat lady had not quite sung yet.

7.

Money, money, money

The Crossrail Bill had been enacted, but there was no guarantee that the scheme would go ahead since it still faced strong opposition both among civil servants in Whitehall and politicians in Parliament. Tortuous negotiations, complicated by shifting political sands in London, had not resolved the matter of funding. While the bill was before Parliament, there was much behind-the-scenes manoeuvring by various players keen to see the project through, but they were frustrated in their efforts to get a definite commitment.

The mood in government, as we have seen, had changed radically. There was now largely a presumption in favour of Crossrail among government ministers. Gordon Brown, chancellor of the exchequer until he became prime minister in June 2007, was amenable to the idea, largely because of the strong support in the City, which he perceived as the economy's key engine of growth. Ken Livingstone, having been re-elected as

London mayor,* this time on a Labour ticket, in 2004, was actively supportive, pushing for the scheme in Whitehall. The only problem, however, was that Livingstone and Brown loathed each other. This was partly a matter of contrasting politics – the two men were at opposite poles of the Labour spectrum, but also, more specifically, because Brown imposed a public–private partnership on the London mayor for the refurbishment of the London Underground. Prime Minister Tony Blair had to intervene numerous times to broker a deal over the core funding. Discussions intensified in 2007 as the Crossrail Bill was being discussed in Parliament and pressure mounted to complete a deal before the May 2008 London mayoral election. The presence of numerous other players, notably the Department of Transport, the City of London, BAA, which owns Heathrow, and Canary Wharf, all with specific agendas, meant that negotiations were highly complex. Even though a PFI deal had been ruled out, Brown was keen to extract as large a contribution as possible from the private sector.

Insiders believe that the key moment was the election wobble in the autumn of 2007. Throughout that summer, Gordon Brown, newly installed at Number 10 following Blair's resignation on 27 June, had allowed his close confidants to hint that there would be an early election – either that autumn or in May 2008. Speculation reached fever pitch by the end of September and Brown was keen to get 'good news' stories into the media. Crossrail, with its appeal both to the City and business

---

* Livingstone served his first term as mayor (2000–04) as an independent, having pushed the official Labour Party candidate, Frank Dobson, into third place behind him and the Conservative candidate, Steve Norris.

interests as well as to Londoners in general, was an obvious candidate and therefore Brown announced on 4 October that, since a deal had been reached with the City of London and the various other players, the scheme could now proceed. Crossrail had struck lucky, for, within days of announcing the scheme's go-ahead as a way of creating a pre-election feel-good factor, Brown had panicked and dropped the idea of holding a snap general election. Although the election never happened, his announcement and commitment made Crossrail far more difficult to cancel.

The funding deal was announced at the same time and published in Parliament the following month.[1] The cost of the project was now put at £15.9bn, including contingency, with the biggest contribution – of just over £5bn – coming from the Department for Transport. The National Audit Office, in a report published in 2014, revealed that 'cost estimation techniques indicated there was a 95 per cent probability that actual costs would be £15.9 billion or less',[2] which is perhaps why ministers were happy to proceed with the project.

'The City of London, in particular,' Gordon Brown said, 'will need to make a significant contribution.' Despite the pretence that much of the money was coming from private sources, it was the taxpayer – in one way or another – who would foot the bill. TfL was committed to contributing a grant of £2.7bn and the Greater London Authority (GLA) a further £3.5bn. However, the latter was a major innovation since the money would be raised from London businesses through an extra rate of 2p in the pound annually between 2010 and 2039 across the capital by business in premises with a rateable value of £55,000 or more (later, following a revaluation, raised in

2018/19 to £70,000) – in other words only large companies. This supplementary rate required new legislation to be passed, which was expected to come into force in 2010. The decision to impose the tax on all businesses throughout London that met these criteria – about one in five of the total – was a politically courageous one: it could well be argued that businesses in Penge or Pinner derived little benefit from Crossrail except – rather vaguely – in terms of London's overall prosperity. Rather than accounting this as an income stream over this long period, the money raised in this way is 'securitized'. This means that all the future revenue is added up and discounted (since a pound next year is worth only around 95p today, and a pound in two years 90p and so on) to give what is called net present value (NVP). In fact, using this calculation, the estimate of £3.5bn has proved to have been rather conservative.

Another major chunk of the funding was guaranteed by Network Rail. (In 2007 it was a quasi-private company, but is now effectively a state-owned corporation and very much on the government's balance sheet.) Its contribution of £2.3bn was to pay for the cost of the improvement and refurbishment of the lines outside the central tunnels that was necessary to enable them to operate the Crossrail services. Network Rail will, of course, recoup its investment by being allowed to impose higher track access charges for those using the improved lines.

One curious aspect of Crossrail's cobbled-together funding package was the £400m deriving from what the document called 'LU interface savings'. When the funding heads of terms were published in 2007, the interface with London Underground was considered to be one of the project areas most at risk of overspending. TfL was asked to guarantee £400m of contingent funding against the potential risk, and this was included as a contribution. However, if TfL managed the risk and ensured that the money was never needed, this £400m of contingent funding could be taken out of the equation.*

Sales of land, developer contributions such as Section 106 payments and the 'London Planning Charge' levied on every major planning approval (which later became the Community Infrastructure Levy – CIL) were expected to raise a further £1bn. Three major private contributions were the subject of intense negotiations. The smallest was the £54m from Berkeley Homes to help pay for the station at Woolwich. Michael Schabas commented that 'the government does seem to have had some success extracting real money from a developer with the threat that without it there would be no station'.[3] However, since Berkeley Homes were planning to build 3,700 homes around the Woolwich station, that represented only around £15,000 each – a figure far short of the value added by the presence of a station that could whisk local people to the West End in just twenty minutes.

BAA, the owners of Heathrow Airport, agreed to pay £230m, though it later transpired that this was subject to approval by the Civil Aviation Authority as the company planned to recoup the money through higher landing charges. BAA, which built the tunnel between the junction on the Great Western line and the airport, also wanted to levy heavy access charges on Crossrail trains for using the track and both BAA's contribution and the

---

* As we shall see in the next chapter, that is exactly what happened in 2010 in the assessment of the scheme for the Comprehensive Spending Review.

issue of track access charges led to lengthy disputes. Martin Buck, a director of Crossrail until 2016, says that Heathrow's total contribution was a mere £70m, the amount the company was allowed to put on its regulated asset base (the RAB, which is the capital sum against which the landing fees they can charge are assessed and levied).*

The most complex negotiations were those with Canary Wharf Group. Their contribution nominally amounted to £150m, which was to be used to construct the station serving the ever-growing complex. However, instead of handing the money over and letting Crossrail contract the work, the owners of Canary Wharf thought up a clever wheeze: *they* would build the station instead and take the cost risk. Crossrail had estimated that it would cost £850m to build the station – around double the expense of most of the others since it was located on a particularly difficult site. Canary Wharf reckoned it could be done for just £500m and were happy to guarantee that sum. No one knows what it cost in the event to build Canary Wharf station. Michael Schabas, however, makes the telling observation that 'the taxpayer got a much better deal than if Crossrail had built it themselves, but Canary Wharf Group seem to have done pretty well too'.[4] Indeed. The station is enormous as it includes four levels of space above the tracks for retail from which Canary Wharf will earn substantial rental payments.

The final piece of the funding jigsaw was a vote by the City of London's Common Council to agree to make a substantial

---

* So, for example, if the RAB is £1,000m, and the rate of return is assessed at 10 per cent, then the company would be able to raise £100m through landing charges and they would be set accordingly.

contribution. It committed to a payment of £250m, but also promised to raise a further £100m in a 'pass the hat' arrangement – though this never materialized because of the impact of the financial crash. The first tranche of £200m was paid over only in 2017, and payment of the remaining £50m was deferred until 2019 as part of the deal which has seen the City fund half of the costs of Crossrail's art programme, detailed in Chapter 10.* Bearing in mind the huge benefits that the City's businesses will derive from the new railway, this hardly amounted to Gordon Brown's 'substantial contribution'.

However, at the same time as providing details of Crossrail's funding, the document published in November 2007 also confirmed the uncertainty about the amounts of money going into the scheme from the private sector: one clause specified that arrangements with the various players needed to be 'finalized'. Overall, only a very small proportion of Crossrail's budget was raised by genuinely private funds: neither the City nor Network Rail can really be considered to be private-sector organizations and the main contributions from the two developers, Canary Wharf and Berkeley Homes, and the airport were all subject to various caveats. On the other hand, Londoners and London businesses are contributing about 60 per cent of the total cost of the scheme through various mechanisms such as the extra business rate and the contribution from the Greater London Authority.

---

* In the words of Crossrail's website: 'Crossrail has… sought to integrate art into the project from the start of construction. Through a range of arts projects varying both in scale and nature, Crossrail is working with both established and emerging artists, international and local, to create an ambitious and diverse art programme that reflects both the Crossrail railway and the city it will eventually serve.'

Even with the funding structure announced and the legislation on the statute books, there remained no shortage of doubters. An article in *Rail* magazine, published soon after the bill was passed, commented:

> Everyone heralds Crossrail as top priority, as vital for London's future and for UK plc's revival. But while the City and business sing Crossrail's praises in public few seem to have any loose change to flip onto the collection plate. Crossrail's £15.9bn funding package looks fragmented, and many of the plans to recoup the money borrowed for construction are still on the drawing board... For such a *grand projet*, it's baffling that the funding regime isn't more trenchant. Those still-under-construction funding mechanisms mustn't stall or deflate any part of Crossrail's project ambitions.[5]

The politics of Crossrail was even more complex than its funding. The passage of the bill did not ensure that the scheme would definitely be built. Yes, the government was making supportive noises; and, yes, the City and Canary Wharf, along with other powerful private interests, were batting for it, but £15.9bn is a lot of money, and there were several moments when it could have gone either way. The politics had been greatly helped by the strengthening of the business case for Crossrail through the ploy of including 'agglomeration benefits'* in the cost–benefit analysis as a component of the 'wider economic benefits'.

Economists argue that such benefits derive from the way that transport schemes improve market efficiency. They might

---

* See Chapter 3, p. 48.

include, for example, the increased potential for development or the fact that people will be able to work longer hours as their travel time is reduced. Essentially, they are a way of calculating the secondary consequences of building new transport infrastructure and are now routinely included in business cases. According to the Colin Buchanan & Partners transport consultancy, which provided a renewed business case for the scheme following the passage of the bill through Parliament, Crossrail would bring in £16bn in 'user benefits'. These were made up of £11bn enjoyed by commuters, tourists and day-trippers and £5bn by business travellers. With the scheme costed at £9bn (without contingency) the benefit–cost ratio was now 1.8:1. But bring in the 'wider economic benefits' such as job creation, the taxes paid during construction and economic growth stimulated by the line, which total between £39bn and £65bn, and the benefit–cost ratio zooms up to between 4 and 5. A no-brainer to fund, surely.

Livingstone and Brown had agreed on the core funding, but Livingstone departed the scene when he lost the 2008 London mayoral election to Boris Johnson, who became the first Conservative to hold the post. Although famously lazy and averse to detail, Johnson was enthusiastic about Crossrail. As one insider put it, 'he did not play games about Crossrail, he backed it and helped keep up the pressure'.[6]

But even with the money in the bag (sort of), not everyone was convinced. Lord Adonis, who took over as transport secretary in June 2009, recalls that keeping Crossrail on track was a constant struggle during his time in the Department for Transport, where he had been a junior minister since October 2008. He says that the battles were constant: 'At every discussion I pushed

Crossrail forward. All I did, week by week, day by day, was to back Crossrail, stop blockages, have rows with Treasury officials and, after we left government in May 2010, tell Philip Hammond [who replaced Adonis as transport secretary] not to flinch.'[7]

There were in fact several occasions when Crossrail could have been ditched or delayed during the period between the passage of the bill and the letting of a series of major contracts at the end of 2010. In the aftermath of the financial crisis that peaked in 2008 and caused a lasting recession, the private-sector players were beginning to baulk at going through with the contributions to which they were committed. In particular, Adonis remembers the Berkeley Homes' chairman seeking to renegotiate his company's contribution: 'All the private investors who were planning to put money in like BAA and Berkeley Homes were threatening to run away. I had to let Tony Pidgley [chairman of Berkeley Homes] off most of his contribution as he threatened to walk away. Yet Berkeley Homes will do fantastically well out of Crossrail and, in truth, we should not have let him off but, rather, asked for two or three times the amount.'

Adonis accepts he was under the cosh from the developers: 'If Berkeley Homes had walked away, that would have meant the station [Woolwich] would not have happened and the officials were saying then that the whole branch to Abbey Wood would not be built.' Adonis was particularly keen to retain the Abbey Wood station as it is in Thamesmead, one of London's least accessible suburbs and in vital need of regeneration: 'I had visited Thamesmead when I was an education minister and it is one of the most deprived council estates in London.'

Adonis provides a fascinating insight into how difficult it was to keep Crossrail going in the face of much internal opposition within government circles: 'There were endless decisions to make to ensure progress was made, such as over the type of signalling, an issue over which there was almost internecine warfare, how much TfL would have to pay, whether we would go to Abbey Wood and so on. Every squabble between the Department and TfL would end up on my desk and I had to sort it. I would make a major decision every week and every one was geared towards keeping the thing going.'[8]

Adonis lost one major argument which republican-minded Londoners will regret. 'I thought it should be called the Churchill Line because Churchill was a great London figure and he saved London. But Boris was intent on Elizabeth Line after chatting to some Royal – I am not sure which one – and the pull of the Royals became too great.'[9]

Crossrail survived the financial crisis, but when Labour lost the May 2010 election and the Conservative–Liberal Democrat coalition was formed, there was a risk the scheme would be scrapped – despite the fact that both the Labour and Conservative parties had committed to the scheme in their manifestos. Much to the promoters' relief, Philip Hammond, the new transport secretary (and future chancellor of the exchequer), confirmed that the coalition government was committed to the project and reiterated that the first trains would run in 2017.

Even at this stage there was no absolute guarantee that the project would go ahead, since none of the major contracts had yet been awarded. Despite Labour's election loss, however, Adonis was confident:

Having a Tory mayor and a Tory-dominated coalition government helped a lot. If there had been a Labour one, they might have been minded to cancel it. Secondly, the way that it was funded made it difficult to cancel once money started being spent. So the Treasury could not play its usual game of spending a small amount of money and then scrapping the scheme, because it was not only their money that was involved. With such a large contribution from London taxpayers, once spending started, it became very difficult to cancel Crossrail as Londoners and London businesses would have demanded a refund, so the money would have been spent twice all for no reward.[10]

Crossrail's future seemed assured. But there was to be one further hiccup.

8.

A daunting task

The money was there, the politics seemed to be favourable and everyone was on board. Now, at last, the Crossrail team could get on with it. How, though, does such a massive project get started? There is so much to do, so much to work out, so much phasing and estimating – and all this accompanied by constant anxiety as to whether the money will come through and the political support will remain in place. There is, however, the comfort of knowing that once a certain amount of work has been done, the project becomes unstoppable. Even costly disasters like the Big Dig in Boston* and the Scottish Parliament tend to get finished. There are relatively few remnants of half-built megaprojects dotted around the world.

Following the passage of the bill, the legal status of Crossrail had to be established. An agreement between the Department

---

* A megaproject in Boston, Massachusetts, which entailed the rerouting of Interstate 93, a motorway running through the heart of the city, into a new 1.5-mile (2.4-km) tunnel, and which overran massively in terms of budget and was delivered more than a decade late.

for Transport and TfL formalized the overall management, ownership and governance of Crossrail. Cross London Rail Links, which became Crossrail, was established as the project delivery agency, and became a wholly owned subsidiary of TfL.

At last work could start. Or, rather, demolition. Almost the first inkling Londoners had of their new railway line was when the Astoria, the iconic music venue on the corner of Tottenham Court Road and New Oxford Street, disappeared in the summer of 2009. The Astoria had hosted its last concert on 14 January 2009, on behalf of Billy Bragg's Jail Guitar Doors charity and the Love Music Hate Racism campaign. The venue was compulsorily purchased by Crossrail and – despite last-minute protests and demonstrations by the Save the Astoria campaign – by the autumn it was no more. The site was needed for the expanding Tottenham Court Road station, one of the most cramped in central London and constrained by the proximity of the huge, thirty-three-storey Centre Point, which itself was in the throes of being changed from office to residential use.

The first construction work took place at Canary Wharf station on 15 May 2009, at a ceremony attended by a pair of mismatched politicians, Prime Minister Gordon Brown and London mayor Boris Johnson. Transport minister Lord Adonis, London's Transport commissioner Peter Hendy, Crossrail chairman Douglas Oakervee and Canary Wharf Ltd chief executive George Iacobescu were also there to watch the first of the nearly 20 metre-long (65.5 ft) steel piles, which underpin the massive station, being driven in. It was a canny move by the owners of Canary Wharf, who were responsible for building the station on the Isle of Dogs. Coming so quickly after the bill was signed, it emphasized the involvement of the private sector – even though, as we have seen, the overall contribution from non-public sources of money was relatively small. It also helped to create an unstoppable momentum.

The need for speed was well demonstrated by the fact that there was still one last wobble over the public funding, even after the preliminary work had started. Crossrail was required under the legislation to publish an annual statement about the project's funding and finances. The statement issued in January 2010 revealed that the estimated cost had escalated from £15.9bn to £17.8bn, provoking a media backlash and a predictable outcry from politicians opposed to the scheme. In response, Crossrail's managers sought desperately to reduce costs and risks. Their efforts became all the more urgent when the new coalition government – obsessed as it was with projecting an aura of fiscal prudence in contrast with the supposedly spendthrift tendencies of its Labour predecessor – announced it would publish a Comprehensive Spending Review in the autumn of 2010 and let it be known that this would include a reassessment of the project.

This was a dangerous period for Crossrail. The amount of work that had been carried out by that stage was minimal and therefore there was no certainty that it could not just be scrapped. That summer was a time of genuine concern in the team. The political zeitgeist was obsessed with 'cutting back the deficit': specific targets were being set for each year and austerity was the watchword. As a big-ticket item, Crossrail could easily be culled, especially since, superficially, it could be argued that it benefited only Londoners. The recession had turned out to be a long one and recovery was slow, though fortunately passenger numbers on the railways and the Underground declined only for a year or so before starting to creep up again. Nevertheless, the usual tropes about how project costs always escalate seemed to be valid in this case, considering that a scheme once costed – albeit unrealistically – at a mere couple of billion was now heading towards ten times that figure.

And there were also attacks on Crossrail from within London government. In February 2010, the London Assembly Transport Committee, which was Tory-dominated but chaired by the Liberal Democrat Caroline Pidgeon, published a report that was broadly supportive of the scheme but which questioned the extent of London's contribution and expressed fears about the potential for huge cost overruns:

> The Committee recognises that London will benefit substantially from the construction of Crossrail. That said though, it is making arguably an unfair contribution to the project's costs. This especially appears to be the case when compared with the contribution made by, and expected benefits accruing to, central government and areas on the route outside London. This generous contribution would become particularly relevant in the event that there were cost overruns and additional funding is required to complete the construction.[1]

In advance of the Comprehensive Spending Review, frantic discussions took place between all the key players: the Department for Transport, the Treasury, TfL and the Crossrail team. One reason for the increased costs was that Network Rail had estimated that it needed an extra £800m to carry out the work beyond the tunnels, pushing their total contribution up to £3.1bn (from £2.3bn). Network Rail has a poor reputation within the industry over cost escalation on its projects and this proposed increase was quickly knocked on the head, with the company being asked to try to spend just £2bn in order to reduce the overall budget.

However, other cuts were needed to save the project as, that summer, the Treasury started agitating for severe reductions in public expenditure. George Osborne, the chancellor of the exchequer, was basing his whole reputation on the ability to bring down the deficit and that could have spelled the death knell for Crossrail. Work was hardly underway and no major contracts had yet been let. When the bill went through Parliament, the design work was only around 30 per cent complete, and while there had been considerable honing of the project's finances in the intervening eighteen months, the cost estimates were still necessarily vague.

Boris Johnson, the London mayor, who was now looking to the mayoral election in 2012, piled on the pressure to save the project. He held a series of meetings with Prime Minister David Cameron and George Osborne, arguing that, if the project were scrapped, it would scupper his chances of being re-elected. 'Savings' of £1.6bn were therefore found to satisfy the Treasury, half of which came from 'reducing risks by simplifying integration works, re-sequencing work and reducing scope, saving £800 million'.[2] This amount included the £400m, mentioned in Chapter 7, that Transport for London had set aside because of potential risks for the Underground's interface with Crossrail; it also included savings from a decision not to create a direct connection from Crossrail to the District and Circle Line platforms of the London Underground at Paddington station, a decision that was later reversed at the instigation of Transport for London.

The 're-sequencing' resulted in the whole project being delayed for a year: the tunnels would now open at the end of 2018 rather than in the spring of 2017 (the date set in the original funding case back in 2007), as this would save money on the tunnelling costs. At the time, it was expected the full running from end to end would not take place until late 2019, but after the postponement announced in August 2018, it is unclear when services will be run along the whole line. The notion that a delay would save money was highly unusual: an extension to the deadline of a major project normally results in extra costs, arising from the need to hire employees and borrow money for longer (as well as delaying the receipt of any revenue). However, Crossrail managed to convince Osborne that the delay would save money through better programme management. In a report published in 2014, the National Audit Office concurred with this view: 'this approach [of delaying the project] reduced the level of risk in the programme and therefore the costs allocated to cover risk'.[3]

The two other sources of savings were even more controversial. Firstly, £300m would be saved because the recession had resulted in lower inflation forecasts; secondly, with the building industry desperate for contracts, it was estimated that £500m could be knocked off the bids put in by contractors. So that was £800m saved, on paper at least. The Crossrail team was effectively gambling on the expectation that, in the context of the highly negative effect of the downturn on the construction trade, contractor prices would be lower than previously calculated. There was an element of game-playing here. The Treasury was probably well aware that these cuts were more theoretical than real, but it was playing to its audience by presenting itself as parsimonious and disciplined in relation to such a big project at a time of austerity. For its part, the Crossrail team also realized that these were mere numbers, which might ultimately be wildly wrong; but they knew that they had to do something to satisfy the bean counters in Whitehall. The agreement between Crossrail and the Treasury did not attract much media interest: there was plenty of other red meat for them in the Comprehensive Spending Review – the first to be undertaken in the new period of austerity by the Tory-dominated administration. Thus the Crossrail scheme received little detailed press scrutiny.

Although to some extent the project was still on trial, the preparatory work on Crossrail continued unabated while these political and financial questions were being resolved. The project delivery agreement between Crossrail and its two sponsors, the Department for Transport and TfL, which was finalized after the Crossrail Act was passed, stipulated that the project had to earn its independence through a series of four review points at which its performance would be assessed. This was, in effect, an insurance policy for the two sponsors to ensure that their money was being spent wisely and that the project was on course.

At each of the four review points, Crossrail assumed more power and independence. The first two review points, establishing the legal framework and structure of the organization, were passed in 2008. The third was a successful interim review of the project in September 2009, which gave Crossrail the power to issue tenders for contracts, though not to sign them off. It was the fourth point, divided into two stages, that ensured its full independence of government. It was not until December 2010 that the first of these final hurdles was passed when the Crossrail board was given the power to let contracts. A few months later, in April 2011, full operational power was transferred to Crossrail or, legally, to TfL, since Crossrail became one of its fully owned subsidiaries.

Terry Morgan, who arrived as chairman of Crossrail in June 2009 straight from Tube Lines, one of the two companies in the failed London Underground private-public partnership, says that he inherited a 'well-thought out plan'. But at this point the organization was in a state of flux and had not yet passed its last two review points: 'We were in a period of transition from planning to delivery and we did not have a lot of authority at that time. We could not allocate a contract, nor did we want to.'[4] When he arrived, virtually no work had been carried out – just a bit of piling: 'We had a couple of hundred people in the development team in Victoria and so we needed to build it up with a big recruitment drive and those who had been involved in planning were basically being eased out and being replaced with people who were going to be responsible for delivery.' Fittingly Crossrail soon moved its HQ to Canary Wharf.

Morgan, who took over from Doug Oakervee, who had been executive chairman for three years and had overseen the parliamentary process, was supposed to be a non-executive but he has taken an active role in the development of the project all the way through. He arrived a couple of months after the key appointment of a chief executive. Given his availability, the choice of Rob Holden was not a surprise: Holden had worked to deliver the Channel Tunnel Rail Link for a decade and was the chief executive who saw it through to completion. He had the right experience and, as an accountant, would be able to keep a close eye on the budget. Keith Berryman, who had been chief executive during the planning stage – and was a key figure in seeing the bill through Parliament – stayed on to oversee the land acquisition. He had been with Crossrail since 2000 and Morgan did not want to lose him: 'Keith was the fount of all knowledge and it was vital to keep him on.'[5]

Morgan and Holden did not want to be rushed into starting the serious work of letting the contracts too quickly. Morgan says: 'We were under pressure to do things before we were ready, but that would have been a mistake. But there was still a question of what sort of governance structure should we have.' This is a vital issue. Big projects like Crossrail can, as Flyvbjerg stresses in his book *Megaprojects and Risk*, get completely out of control. There is no shortage of examples, including two notable ones from Germany: the cost of the reconstruction of Berlin Brandenburg Airport soared from around €2bn to €7bn and the opening delayed for almost ten years by a series of engineering issues, including a faulty ventilation system (it is apparently designed to blow smoke downwards, rather than upwards, in the event of a fire and not surprisingly this revolutionary concept has caused no end of problems) and escalators that are too short, as well as by political wrangling between the two public authorities involved; and Stuttgart 21, like Crossrail a tunnel under the city linking two railway lines, where costs have almost doubled to €8.2bn and the project completion delayed by eight years. Closer to home, the Channel Tunnel's failure to keep control of its contractors saw a rise in costs of 80 per cent to £4.65bn. Such cases are ever-present in the minds of Crossrail's senior executives as a reminder that when a mega-project goes wrong, it can go very badly wrong. Even with the large contingency available, there is a risk that things will still go awry.

Bechtel, a big US project management company, was appointed by Crossrail as the project delivery partner and its team was headed by Bill Tucker, who arrived in April 2009. Tucker's expertise was power generation, but he had cut his teeth on the railways as a member of the Bechtel team called in to rescue the West Coast modernization programme, a project which had gone badly wrong and was the main contributing factor in the collapse of Railtrack into administration in 2001. Terry Morgan took the key early decision that there would only be one team, despite the fact that Crossrail did not directly employ many of those, like Tucker, working in Canary Wharf: 'I did not want two teams in one business. I wanted one team. If we knew we wanted someone for the whole ten years, then we would take them in-house. But we also had many people from Bechtel and that is how we got the best out of them. It was a melded team.' Once a contractor was taken on, Morgan's vision was that their people were working for Crossrail: 'Inside the business you could not tell who was doing what. Everyone was part of the Crossrail team and that was clear from the outset.'[6]

Touring the various sites with Morgan, it is clear that there is no difference in how he interacts with direct employees or contractors. Creating that team spirit, he reckons, has been a key part of the success of the project in keeping largely to time and budget. There was, however, a change of plan regarding management of the work on the central section of the tunnel. Initially Bechtel was going to do the whole of this on its own, but Crossrail decided to use some of its own people as well. This saved on cost as Bechtel does not sell its services cheaply. Tucker explains: 'We estimate what staffing we need and we do that each year. We get reimbursed per staff member and we make incentive fees based on performance' Tucker reels off the scope of his task: 'I am responsible for the central tunnel section with nine stations and about £7bn of the total. We spent about £120m every four weeks and now [February 2018] we have twenty-five contractors but there were a hundred and fifty over time.'

When Tucker joined, he reckons that about 30 per cent of the design work had been completed:

> the level of engineering required to get through the hybrid bill process is high. It gives you enough to show the stakeholders with certainty what you need to build the scheme and to demonstrate the viability of the project. We had schematic plans for the stations and the tunnel alignments in that first 30 per cent, which allowed us to develop the overall cost estimates. So our first task was to let contracts for the design of the next level of detail for tunnels and stations.[7]

Reaching a more precise estimate of the costs involved is crucial: 'Some parts you estimate by square metre, others by the amount of material you know it will require, such as metres of cabling and the like.'

While Bechtel was the delivery partner, Crossrail also appointed a programme partner. In March 2009 it chose Transcend – a joint venture between AECOM, CH2M Hill and Nichols Group – for the task of ensuring the smooth running of the programme. The sheer scale of what Transcend had undertaken is demonstrated by the consortium's announcement in early 2017 that it had clocked up a million staff hours – the equivalent of employing 500 people for a year – in time spent managing the correct phasing of all the various aspects of the work.

John Pelton, a director of Transcend, who later became the strategic projects director for Crossrail, explained how the system works: 'Crossrail hires the key people and the people it thinks it needs in continuity jobs particularly, then the programme partner either brings the programme skills that are difficult to find from open recruitment, or provides the ability to respond to specific challenges, short-term problems, and technical issues – whatever it might be.'[8] Like Morgan, he emphasized the critical importance of everyone at Crossrail working as a team, with no demarcations between the various contractors involved. This greatly facilitated Transcend's role of co-ordinating the work of everyone on the project.

Simon Wright, the programme director and later chief executive during the final stages of the project, outlined the managerial complexity of the programme: 'There were probably a hundred key dates, which all feed into opening the railway on 9 December 2018 [the day the main tunnels were supposed to open for business]. We work back from these and there's probably a thousand pages of schedules feeding into that.' On the day I spoke to him, in March 2018, he was able to tell me, with impressive precision, that the scheme was 91 per cent complete. That misplaced confidence in a metric that was clearly wrong would prove to be the Crossrail team's undoing.

The management structure of Crossrail was based initially on the three key major tasks: tunnels, portals (entrances) and shafts; stations; and rail systems. During the tunnelling phase, this structure changed to an east–west arrangement based on geography rather than function, then reverted to the original structure when the tunnelling was complete but Bill Tucker retained overall responsibility throughout. It was clear that the first contracts had to be to dig the tunnels, as only a limited amount of work could be done before they were available to use as access to the various station sites. Before contracts could be let, Crossrail had to 'earn the right to run the business', as Terry Morgan puts it. In other words, the Crossrail team had to show it was capable of letting and managing these massive contracts within what, at the time, was a £15.9bn budget.

As soon as Tucker arrived at Crossrail, he was keen to establish a rigorous safety regime. Construction is one of the world's most dangerous industries and tunnelling is a particularly risky aspect of it. During preliminary discussions about the building of the Channel Tunnel, a consultant's report suggested that there would be twenty-four deaths. In the event, ten workers, eight of them British, lost their lives during the construction of the Channel Tunnel, most in the first few months of boring, but the Crossrail team considered this unacceptable and adopted, right from the start, a Target Zero approach.

Safety in relation to both workforce and public was fundamental to the philosophy of the project and its executives focused on this point in the many conversations I had with them. Simon Wright, Crossrail's chief executive and programme director, explained that risks have to be minimized in a way that is actually more restrictive than the public's day-to-day experience of public transport: 'Human beings take unbelievable risks and in construction you can't do that. Take the Tube. Everyone will happily stand on the platform edge of the Northern Line with a train rushing past them at thirty mph and the risk is huge. On a project like this, you cannot take that sort of risk and we have to persuade people of that... It is called behavioural safety – persuading people to take care of themselves is a key issue.'

The concept of behavioural safety is based on three principles: 'we all have the right to go home unharmed every day; we believe that all harm is preventable; and we must all work together to achieve this.' These may seem glaringly obvious truths, but traditionally the building trade has been cavalier about health and safety, and the high casualty rates it has tolerated are unacceptable in the twenty-first century. Crossrail set out a series of 'golden rules' to back up these principles, such as respecting the basics, assessing the risks and supporting each other. Much has changed in terms of safety from the time when my friend Liam Browne, as a thirteen-year-old, was whisked down to the construction site of the Victoria Line in 1967 in a bucket attached to a crane as part of a Saturday job:

My first day was a cold, dark, frosty morning and I followed my dad up a fifteen-foot ladder to get over the cast-iron rings of the shaft head and then, looked down a hundred and twenty feet to the bottom of the wide shaft. Five long ladders, roped to the side of the shaft on scaffolding platforms, took me out of the cold air and into the warmth of the tunnels. No hard hats or special gear, just men wearing their old suits. Occasionally, when the crane driver was in, we would get a lift down the shaft in the huge cast-iron bucket used for bringing up the clay. The crane driver would lift it up about twenty-five feet in the air over the rings and then it would drop at a terrifying speed, to about ten feet from the bottom where it slowed. On the way back up, at the end of a shift, men would run to get the bucket lift to save climbing up all those ladders and to get an extra ten minutes in the pub, so some would even hang onto the outside of the bucket or hurl

themselves at it as it began to take off, holding on precari-
ously for the ride.

On Saturdays the shield was not operated, unlike at Crossrail
where it was a 24/7 operation, but one day Browne saw it working: 'The miners were in because they had got behind during the
week. These were hard men who earned big money, got paid on
Friday and had usually drunk it all by Tuesday.' In some ways,
the system was the same as today and its roots in Greathead's
concept* are clear, but safety was barely a consideration:

> The shield was a large circular ring with a platform across
> the diameter where three men could work on the top and
> three on the bottom, digging out the clay with pneumatic
> drills. The shield would be pushed forward and when a metre
> was completed, cast-iron ring segments would be lifted into
> place. These were brought to the shield on bogies and lifted
> by winch at which point the miners were supposed to stop
> drilling while the rings were put in place, but these men were
> always in a hurry. The ring was only attached to the lifting
> wire from one end and so it would swing wildly when it
> lifted off the bogey. The miner nearest the sidewall, one eye
> on his work in front, the other on the ring, would weave
> casually to avoid the two tons of cast iron that was practi-
> cally brushing his shoulder and which could easily be deadly
> if it hit him.

Those days are long gone. Men no longer work with pneumatic

---

* See Chapter 1, p. 10.

drills at the face of the boring machine – the work is now carried out by automatic cutters and scrapers. And there are certainly no men hanging on the outside of buckets being hoisted up the shaft. Safety briefings are de rigueur, even for casual visitors, and rules about wearing hard hats and the garish but highly visible fluorescent PPE (personal protective equipment) are rigidly enforced. Everyone on site has to wear this equipment, including solid boots and protective glasses, when certain types of work are undertaken. Safety messages are posted everywhere. No untrained teenage sons of the workers are allowed in to do a Saturday shift. If these new procedures can feel process-driven and tick-box in nature, they have undoubtedly contributed much to the reduction in casualties in what remains one of the world's most dangerous industries. Unfortunately, not all of the contractors managed to live up the high standards expected by Crossrail: as we shall see, the death of one worker and serious injuries to two others resulted in the prosecution of the companies involved.* Nevertheless, Crossrail can claim a relatively good safety record for such a vast project.

The first stage of delivering the project was to fill in the details of the design. Since this was far too big a task to carry out in-house, a series of twenty-one design contracts were let, relating to particular stations or features, such as the tunnels and the shafts and the communications and control systems. The pool of contractors capable of taking on this type of work is not large. In the end Mott MacDonald, an engineering, management and development consultancy, won nine of them, with the remainder split between half a dozen other companies. The

---

* See Chapter 10, p. 228.

task of these 'framework design consultants' was to bring the design down to the level of detail required to let the construction contracts. In technical terms, they moved from the Royal Institute of British Architects Stage D, the scheme design, to Stage E, which is the technical design. Crossrail, Tucker states, 'retained overall responsibility for managing the design and ensuring the integration of the design across all the various interface points'.[9] In other words the big decisions were made in-house by Crossrail.

Since most of the stations could not be built until the tunnel boring machines had excavated the running tunnels, contracts for that work were the first major engineering jobs to be let. The four tunnel boring machine contracts were put out for tender in the summer of 2009 and awarded by the end of 2010, after the Comprehensive Spending Review revision. Although by then a huge amount of work had already gone into preparing the way for the tunnel boring machines and some preliminary work had been undertaken at stations, the letting of these contracts really marked the beginning of the project. Crossrail was by then unstoppable.

9.

Digging under London

Starting the big dig under London had required a long period of preparation. Decisions had to be made on the logistics of how many machines were required, where they could be assembled and what sections they would excavate. The first task, however, had been to establish in great detail the geological conditions along the whole route. Quite a lot is known about what lies beneath London, but before the project started there were also many 'unknown unknowns'. As Mike Black, Crossrail's head of geotechnics, has observed: 'understanding of the geological structure of certain areas has increased significantly as a result of the Crossrail works.'[1]

Secondly, there was the potential for obstructions and each separate risk from these had to be assessed. As we have seen, the work of the early Crossrail team had resulted in an alignment that avoided any major structures (even secret American military installations). During the preparatory stage, Crossrail undertook an extensive desk-based survey of maps and other information sources, to establish what obstructions were likely to be found along the alignment of the route. These might include wells, old boreholes, or the foundations or remains of long-demolished buildings. Of course, the location of many objects was unknown and it was inevitable that some unforeseen obstructions would be found.

There were also some more obscure concerns. During the passage of the bill through Parliament, the Conservative Lord James of Blackheath had repeatedly raised the possibility of an anthrax epidemic being triggered by the excavations. He told the House of Lords select committee that '10 Hayne Street [the site of a compulsory purchase order for an empty car park] may be the site of the missing anthrax burial ground that has been lost for 488 years, since Christ Church, Spitalfields and the church at Charterhouse Square refused to accept the bodies of 282 victims of the anthrax outbreak that wiped out the whole population of Hayne Street at that time'.[2] Rather more likely – and far more dangerous – was the possibility of encountering unexploded ordnance from the Second World War.

In the west there were fewer such risks. Paddington is a relatively recently inhabited area of London, part of the capital's rapid nineteenth-century westward expansion. Here there is little likelihood of coming across unexpected foundations or the remnants of a Roman settlement, or indeed unexploded bombs – for west London which, though receiving its share of bombs, was rather less of a target for the Luftwaffe than the docks in the east. The foundations of the few farmhouses and barns that might be found here would not pose any problem for the tunnellers and, in any case, it would be sheer bad luck if they were on the path of the railway.

As the tunnelling approached the City, however, the risks

became far greater. Human beings have lived here for 10,000 years and this was the site of the substantial Roman town of Londinium. As Gillian Tindall elegantly puts it in her book about the Crossrail route:

> in the City, the old heart of London, and in other long-inhabited places such as St Giles-in-the-Fields, the present-day buildings sit on twenty or more feet of the compacted, churned, reused debris of past habitation. Under soaring modern glass and metal constructions lie figuratively, and sometimes actually, the fragmented bricks of Victorian office blocks and warehouses, eighteenth-century stuccoed terraces, the stones or timbers of earlier houses on the site whether modest cottages or substantial manor houses and below again perhaps a Roman watercourse, shards of a sarcophagus or a tessellated pavement.[3]

All of these could cause problems for the mighty tunnel boring machines.

Quite apart from the possible presence of unexploded ordnance and plague bacteria, it was the fact of not knowing quite what was down there that was troubling for the tunnellers. While there is a substantial body of knowledge about the geological strata below the capital city, there are known to be numerous undiscovered faults and deviations even in such a heavily built-up area. The Crossrail tunnels and underground stations 'are located wholly within the London basin, a large, east-west trending syncline that sits between the high ground of the Chilterns to the north-west of London and the North Downs to the south of the city'.[4] A syncline is essentially an underground bowl-shaped or convex formation of strata which have been laid down over millions of years. Knowing the precise composition of each stratum is critical, since each requires a different method of excavation to tunnel through them.

The second attempt to design the Crossrail route was helped by the fact that extensive ground investigations had been carried out for the early 1990s version of the project. Once Crossrail was relaunched, a further series of ground investigations was necessary because the alignment had been changed in several places. Overall, during both phases of Crossrail, nearly 2,000 exploratory holes were bored, each averaging around 20 metres in depth. Unfortunately, the long delay in the project meant that many of the holes bored in the 1990s – most of them on public highways, which are regularly resurfaced – had been filled in during the interval between the two schemes and therefore had to be re-bored (a classic example of how delaying projects results in wasted expenditure). Digging these boreholes is not without risk: one of the worst injuries during construction was caused when a worker struck a high-voltage electric cable while digging an inspection pit in February 2008 for a borehole at the Hanover Street site where the new entrance to the Bond Street station will be. The man was lucky to survive as he suffered 60 per cent burns. The contractor, Fugro,* was later fined £55,000 and made to pay £30,000 costs, following a prosecution by the Health and Safety Executive for failing to take sufficient precautions and not putting

---

* A Dutch multinational provider of geotechnical, subsea and geoscience services for energy, telecommunications and infrastructure companies.

markings down on the road to indicate where cables had been located.*

Building tunnels under London may be more straightforward than it is in some other cities, but the conventional wisdom that all that lies beneath the city streets is nice solid London clay is sadly mistaken. The ground beneath the western section from Paddington is, indeed, clay and since it contains little water and is of consistent density, provides ideal conditions for tunnelling. East of Fisher Street, an access site used by Crossrail near the British Museum, the geology changes as the tunnels descend eastwards towards Farringdon into a stratum called the Lambeth Group, a complex mix of sands, silts and clays whose unpredictable levels create far harder conditions for tunnelling.

To make matters even more complicated, the tunnels between Farringdon and Canary Wharf criss-cross between the London clay and the various deposits of the Lambeth Group, which are around fifty-five million years old and 'can be particularly challenging for tunnelling and sub-surface excavation due to the variability of material types ranging from stiff clays through to loose sands and gravels, hard limestone layers and shelly beds'.[5] In conditions such as these, unexpected water flows and even instability during excavation constitute a significant danger to tunnellers. Beyond Canary Wharf, the tunnels occasionally went into a stratum below the Lambeth Group, known as the Thanet Sand Formation (whose composition is clear from its name). South of the river is mostly chalk.

---

* This may be enlightening for those Londoners who have wondered about the purpose of the coloured lines of paint frequently seen on roads and pavements where excavations are planned.

The biggest challenge for any major tunnelling scheme is groundwater and the associated threat of flooding. There is also a longer-term issue relating to the presence of groundwater: the permanent structures need to be designed to ensure that the water is kept out. Before the tunnels can be designed and built, it is essential to know the precise source of water pressure and the extent of any permanent flows. During preparations for the first version of the scheme, detailed surveys were carried out on the geology of the area to be excavated in order to minimize the risk of delays caused by unexpected ground conditions. But when more detailed investigations were carried out for the second version of the project, they revealed areas of sand with higher concentrations of water than had been anticipated, posing greater problems for the tunnel designers and greater risk for the tunnellers.

Alongside the dangers to workers of underground flooding or collapse, the boring of these tunnels posed a risk of – possibly catastrophic – damage to the buildings above. Again the sheer numbers are impressive. Above the alignment of the tunnel were a staggering 17,500 buildings that could potentially be affected. For each building, the risks had to be assessed and concerns allayed. More than 65,000 sensors were used to monitor the effects of tunnelling. Nearly 500 arrangements relating to subsidence that might result were made with the occupants of properties potentially affected, in which Crossrail agreed to monitor and make good any cracks. Crossrail's neighbours also had to be satisfied that they would be protected from electro-magnetic interference from the railway, which might affect delicate electronic equipment, a not inconsiderable risk.

The sensors were installed on thousands of buildings across

London, and calibrated to measure the slightest movement within a 100-metre radius of any area being excavated. Above ground, the main method of detecting movement was the use of a sensor 'total station' – or robotic theodolites. Each one focuses light from lasers on hundreds or even thousands of prisms attached to buildings around the site to detect any change. Other types of sensor were used below ground to ensure that the overall tolerance was never exceeded.

As well as the 42 kilometres (26 miles) of main tunnels, a further 14 kilometres of passageways, walkways, shafts and connecting corridors – including around 5 kilometres of temporary access tunnels – had to be excavated using the spray concrete lining technique.* Using sprayed concrete, rather than conventional iron or concrete curved sections, allows greater flexibility in the shape and size of the spaces. Hugh Pearman, author of a book on the design of Crossrail, has written that 'this led to some thoroughly romantic designs, with architects suggesting domed or egg-shaped concourses decorated like stage sets with everything from tropical forests and soaring angels to ornate candelabra'.[6]

The planning of the tunnel boring operation was extraordinarily complex, and it was rendered yet more so by the fact that difficult decisions made early in the process might have to be changed in the light of circumstances. First, the routes and timings of the run of each of the boring machines had to be determined. Simply running two machines through the whole 21 kilometres of tunnelling would have been impossible because of the length of time it would have taken and, in any case, the

---

* This is explained in detail in the next chapter (see p. 212).

various ground-level sections in east London would have meant bringing them to the surface, something that is very difficult to do with tunnel boring machines (TBMs). In addition, a section of the line used an existing tunnel which required refurbishing rather than boring from scratch. Crossrail's managers therefore needed to consider what was the optimum number of TBMs and what was most efficient route for them to follow. Two related questions were where would they start and how would they be disposed of once their work was completed.

The initial idea was that the space for stations and the section of running tunnels in them would be excavated first from above and when the TBMs arrived they would be hauled through the station, since they cannot operate in open spaces under their own power, This would have saved time, according to Bill Tucker, who was in charge of the construction of the central section: 'this means you can get the civils [civil engineering] work done early while the tunnel boring machines are being manufactured and getting set up.'[7] Additionally, the time that the machines are in the stations provides an opportunity for the boring machines to be serviced and cleaned up, something that cannot be done in the confines of the tunnel.

However, at Bond Street, Tottenham Court Road and Farringdon, a different procedure was followed. Bill Tucker explains: 'The western tunnels contractor came to us and said we would prefer to use the tunnel boring machines as a pilot first, and then create the tunnels along the station platforms from the pilot tunnel they had made, digging outwards from the tunnel after breaking through the concrete rings that are laid by the boring machine. They thought it was a better way of achieving the result safely and economically.' According to the contractor, at Tottenham Court Road 'this reduced the programme by seven months, produced savings of £80 million, reduced settlement by 30 per cent and reduced the number of lorry movements by 50,000'.[8] Decisions such as this, taken early in the life of a project, have a considerable impact; and the fact that Crossrail accepted such a fundamental change demonstrates that the team was genuinely open to innovation originating with the contractors.

The machines themselves are awesome – an overused word but accurate in this context. When I saw the tunnel boring machine in action at Canary Wharf, the overriding impression was of its complexity and size, and the small size of the team – just twenty people – operating it. The raw figures are, as ever, impressive: the machines are 150 metres long and weigh more than 1,000 tonnes. In order better to illustrate their size, the Crossrail PR team use the iconic red double-decker London buses as units of measurement which reveal that their tunnel boring machines are ten buses long and a massive 143 buses in weight.

Those Londoners who may have been fooled into thinking that the name 'Elizabeth Line' implies this is just another Underground service will receive quite a surprise when they first venture onto the new service. The tunnels the machines have created are 6.2 metres in diameter, far larger than the standard Tube line diameter of just over 3.5 metres. Not only are the trains far larger – basically the same size as commuter trains running around the network – but there is even space for a walkway along the whole length of the tunnels for use in emergencies.

The machines, which cost £10m apiece, are complex organisms. They not only carve out the tunnel but also immediately put into place the concrete rings that line it. They are perhaps misnamed, since they are not just 'tunnel boring' machines but 'tunnel building' machines, mini mobile factories that leave behind a completed tunnel which then 'only' requires building up the formation including any drainage provision, laying the sleepers and then fitting out with the rails, overhead electrical equipment and other systems needed to create the new railway. Interestingly, the length of time taken by the boring and by the fitting-out was around the same, with the latter being slightly longer, which is a demonstration of the sophistication of modern railways. In Victorian times, once the civil engineering such as tunnels and embankments was completed, it did not take long to complete the railway and run trains on it.

Each machine has a rotating cutterhead at the front (a modern version of the Greathead Shield, and a series of trailers or gantries behind, housing a vast array of mechanical and electrical equipment, including conveyor belts to remove the earth. This necessarily over-simplified description hardly does justice to their sophistication and ingenuity. Not only can these machines propel themselves forward with pinpoint accuracy, they can even, as we shall see, weigh every gram of soil dislodged by the cutters.

The machines were operated by 'tunnel gangs' comprising a dozen people on the machine itself and eight working at the rear of the machine and above ground. These gangs worked in twelve-hour shifts, tunnelling twenty-four hours a day, seven days a week. Their skills are in short supply and nearly all the workers, often dubbed 'miners', came from EU countries, especially Portugal. These are highly technical and difficult jobs. Many members of the Crossrail gangs had worked on other tunnelling projects around the world, while others were trained by these experienced members of the team. Safety was, as ever, a key priority. The Crossrail TBMs were the first in the world to include refuges in case there was a gas leak or fire. They could accommodate the whole team of twenty people and had sufficient air in canisters to last twenty-four hours in the event of an accident preventing their rapid rescue. They were never used apart from in drills.

Ventilation was an ongoing problem and risk, particularly in the upper decks of the gantries behind the cutting machine. The confined nature of the machines meant that fresh air was often blocked or diverted by the structure of the machine and modifications had to be made to the TBM ventilation system during construction to improve the situation.

The cutting wheel consists of disc cutters made of incredibly hard tungsten carbide and scraping tools that are pressed against the tunnel face by hydraulic cylinders. The loosened material drops down and then is raised onto a belt by a screw conveyor – think of a large corkscrew working upwards. The belt continues all the way back to the shaft or portal where the machine started its tunnelling and where the spoil is transferred to a train, barge or lorries.

The tunnel face is continuously monitored by pressure sensors that regulate the turning rate and power of the cutterhead. This is crucial. The machine must not bore too deeply and those checking the functioning of the machine have to know precisely how much material is being excavated. The machine is equipped with a laser system that enables the tunnelling teams to guide the forward progress of the machine with an extraordinary level of precision, accurate to within 5 cm of where they were programmed to be. It is this accuracy that gave the Crossrail team the confidence to carve out the tunnels so close to existing structures, making sure no damage was caused to any of the many iconic and expensive buildings under which they passed.

Then, perhaps oddly to those unfamiliar with the workings of these huge machines, the key aspect that needs to be measured, with enormous precision, is how much spoil is coming from the face of the machine. Two very sensitive 'belt weighers' were used to measure the weight of the spoil: these not only had an accuracy of 0.5 per cent but backed each other up to ensure there was no discrepancy. Mike Black explains the reason for this requirement:

> For every ring advance [about 1.5 metres] made by the TBM [tunnel boring machine] there is a predicted volume and mass of material to be excavated, based on the geology through which the tunnel is being bored. The belt weighers measure the mass of the soil passing along the conveyor to confirm that it aligns with the predictions. This will confirm that there has been no over-excavation at the face, which could lead to voids behind the tunnel linings or ground stability issues leading to excessive settlement at the surface.[9]

The Crossrail team was anxious to avoid any incident such as the one that occurred when the nearby Channel Tunnel Rail Link was being built a decade earlier and a garden, including a shed in which champagne was being stored for a wedding, disappeared into a vast hole during the passage below of a tunnel boring machine. Several houses were also damaged by the collapse, which was attributed to the unexpected presence of a well, and caused a two-month delay to the project.

One of the principal ways of minimizing ground settlement was through 'compensation grouting'. Small pipes called 'tubes à manchette' (or tams) from a specially created shaft were bored into areas of the ground where there was concern about possible settlement. Grout – a cement mix – was then injected under pressure into the tams. This method can be used to stabilize the foundations of a building or a part of the ground where there is sand, gravel or too much groundwater to ensure stability. By lifting the ground slightly, the grout helps to reduce any settlement that may occur at the surface. This method was used extensively at several station sites including Farringdon, Tottenham Court Road and Bond Street, but it was at Liverpool Street, as we shall see, that there were the greatest difficulties. The average settlement across the whole project was, thanks to the mitigation measures, on average just 2 cm, far less than the assumption, which had been 7 cm.

Once sufficient progress had been made to allow the concrete rings that line the tunnel to be placed into position, the cutting wheel and the whole machine were halted while the rings were installed by a hoist just behind the face. Seven rings, each weighing 3 tonnes, and a keystone were fitted at every stage.

Just to complicate matters, the tunnels are rarely straight, except at stations, and therefore the concrete rings differ marginally from one another as they had to be adapted to the curvature of the railway, which meant that the right ones had to be brought to the machine in the correct order. The rings were delivered to the front of the machine by a small 90-cm-gauge railway, which was built with the same hoist that placed the concrete rings as the machine moved forward. This line was then used to transport equipment and materials to the machine.

As the tunnel extended further, junctions were created in order to allow more than one small train to operate.

After much consideration it was decided to have eight tunnel boring machines to undertake ten runs, which meant that two were used twice, having had to be disassembled and then reassembled at another site. Such vast machines cannot be transported from a factory and have to be assembled near their starting points. This caused particular difficulties for the first two to be deployed, Phyllis and Ada. Their starting point was Westbourne Park, just west of Paddington, the busy 'throat' of Great Western's London terminus. Because of fears that the use of heavy lifting equipment near the railway lines might prove dangerous, the two machines were assembled half a kilometre away from the portal where they would start work and then moved on low loaders running on temporary rails to the portal which had been partly excavated in preparation. A small section of tunnel was cut out in advance to prepare the way for each of the TBMs whose work then started in earnest, boring the separate eastbound and westbound tunnels between Paddington and Farringdon.

Phyllis and Ada were named after two rather different women: the painter and writer Phyllis Pearsall (1906–96), devised the A–Z street map of London in the 1930s (supposedly being inspired to do so after getting lost in London while using the latest street map, which turned out to be hopelessly out of date); Ada Lovelace (1815–52) was an early computer scientist (and Lord Byron's only legitimate daughter). The tradition in the tunnelling business that boring machines have to be named after women to ensure good luck dates back to the sixteenth century, when miners prayed to St Barbara, a third-century martyr murdered by her father for becoming a Christian. Apparently God exacted his revenge on the cruel parent who was soon struck dead by lightning, and hence praying to Barbara was seen as a form of protection from lightning flashes and explosions in dark tunnels.

Phyllis started work first, after the inevitable ceremony featuring Boris Johnson – fresh from his re-election as mayor – and the transport secretary Justine Greening, on 4 May 2012, and Ada followed three months later. Both spent around seventeen months on their task, each excavating a 6.8-kilometre-long tunnel to reach Farringdon, with Phyllis completing its journey in November 2013 and Ada finishing its section of the tunnel in January 2014.

Meanwhile, the last two months of 2012 saw the launch of Elizabeth and Victoria (no prizes for the provenance of their names) from a shaft in the Limmo Peninsula* near Canning Town in east London. They were heading for Farringdon, 8.3 kilometres away, which meant the two queens had by far the longest run. They would take thirty months to complete their journey and would be the last to finish, completing their run in May 2015.

This time, with no space to assemble the machines above ground, only the front shields were completed on the surface and then lowered 40 metres into the vast Limmo shaft where the rest of the machine was assembled. Getting the machines started is always difficult since they power themselves by pressing themselves against the rings they have just laid. No rings, no power. In this case, the shields were shoved forward by

---

* The Limmo Peninsula is on the east bank of Bow Creek, close to where the River Lea flows into the Thames.

a machine on temporary greased rails in a short tunnel excavated using the sprayed concrete lining method while the rest of the machine to the rear was assembled. The machines were then launched using a temporary frame to push themselves against.

The third pair of tunnelling machines, Jessie and Ellie (named after Jessica Ennis-Hill and Ellie Simmonds, gold medallists in the 2012 London Olympic and Paralympic Games respectively) made two shorter runs. They were launched from Pudding Mill Lane, where they were assembled on the surface, and tunnelled through to the junction between the two eastern branches of Crossrail at Stepney Green. They did not operate in parallel, with the result that Jessie covered the 2.7 km between August 2013 and the following February, while Ellie, launched in the month Jessie completed its task, finished its run in June 2014.

The Stepney Green junction, where the machines ended up, is one of the most remarkable underground sections of the railway, as it had to be big enough – 50 metres long and nearly 17 metres high – to accommodate both of the TBMs whose runs ended there. (Photographs showing Crossrail's orange army standing in this enormous cavern beneath east London provide an emblematic image of the project.) However, despite the cavern's vast size, it would still have been impossible to dismantle both Jessie and Ellie there simultaneously. This was the reason behind the machines' staggered departure from Pudding Mill Lane – they had to be scheduled to arrive at Stepney Green at different times.

After being dismantled, Jessie and Ellie were lifted out of the shaft and transported by road from Stepney Green to the Limmo Peninsula, where they were relaunched to drive the

0.9-kilometre tunnels from Limmo eastwards to the Victoria Dock portal, from where the line runs briefly on the surface until it reaches the Connaught Tunnel. These short runs took just a few weeks and the work was complete by October 2014.

All six of these tunnel boring machines – made by the German company Herrenknecht AG, as there is no British manufacturer – were of a type known as earth pressure balance TBMs. This type derives its name from the process whereby the pressure of the cutting tools at the front is regulated by the amount of excavated material picked up by the screw-shaped conveyor and, as described above, weighed very accurately. The right amount of pressure must be applied to the rotating disc to avoid collapse – the key risk in any tunnelling process. Therefore the measurement of the pressure being applied has to be extremely accurate. If the boring machine is bringing down too much material and the tunnel is becoming unstable, the machine uses the excavated material to shore up the tunnel lining to prevent collapse. Additives such as bentonite (a type of clay mixed with water) can be injected ahead of the face to increase the stability of the ground that is being extracted.

The remaining two machines, also made by Herrenknecht AG, were called Sophia and Mary (not queens this time, but the wives of Brunel *père et fils*, Isambard and his father Marc, both of whom worked on the first tunnel to be built under the Thames between Rotherhithe and Wapping) and were of the mixed-shield, or 'slurry', type of TBM. These machines function in a different way from earth pressure balance TBMs. They are used in more challenging ground conditions where there is sand or gravel, or where there is high water pressure, which means the tunnel lining cannot be self-supporting even for the short period before the rings are fitted. The cutterhead of a mixed-shield TBM operates in a chamber, kept under pressure by the injection of bentonite, where it forms a more stable surface for the excavated tunnel. These two machines excavated the 2.9-kilometre twin tunnels under the Thames between January 2013, when Sophia started work, until May 2014, when Mary's run was completed.

During the tunnel boring process, a series of nineteen tunnels linking the two tracks were built at gaps of around 500 metres for use by pedestrians in an emergency. As these are much smaller than the train tunnels, the TBMs could not be used to build them; instead they were excavated and then completed using either sprayed concrete lining or steel ring segments, depending upon location and geology.

While the process of digging through tunnels may sound relatively simple, the wide variations in performance between the machines, as well as day-to-day, demonstrate the complexity of the task. There are numerous difficulties to be encountered, ranging from machine breakdowns and other technical issues to unexpected ground conditions and safety concerns about potential ground movement. Crucial to progress, too, was the skill levels of the different tunnel boring crews which could greatly impact performance.

Delivery of the wrong set of rings, for example, or the failure of a conveyor belt, can result in the machines having to stop work. Although the teams operated the machines 24/7, with two or three shifts per day depending on the contractor, there were frequent lengthy stoppages, some of them scheduled, some not. Less than half the time spent below ground was actually productive. On average, the machines progressed 38 metres per day, the top performance of 72 metres in a day being achieved by Ellie on 16 April 2014 between Pudding Mill Lane and Stepney Green.

The maximum number of rings laid per day by each machine varied considerably, too, from Sophie's mere sixteen to Victoria's fifty-nine. Yet despite holding the record for the best day's ring-laying performance, Victoria's tunnelling time amounted to just 402 days out of the 895 spent below ground. The biggest problem the machines faced was the inability of the conveyor systems to cope with spoil from the excavations. According to a study of the machines' overall performance, 'the conveyor systems suffered from several failures, including structural issues. Conveyor designers must be made aware of the clogging potential of London Clay... and provide recommendations for planned preventative maintenance of the system and all its components.'[10] Nonetheless, the machines finished their work within a month of the original schedule, but that was a result of building in considerable contingency and time for maintenance and repairs.

The skills and experience of the teams proved crucial. The later drives tended to be more productive because, as an article in the *New Civil Engineer* explained, 'they were short drives, mined by teams who had already worked together on previous Crossrail drives and who had already been through the learning curve'.[11] The nature of the tunnelling work was encapsulated by Peter Main, the engineer in charge of the drives in the eastern tunnels, and who was filmed standing in front of Ellie after its run from Pudding Mill Lane: 'It is testimony to the professionalism of everyone involved. These things happen not by accident but by design, by strong management, by collegiate working, by trying to maintain a degree of empathy with the contractor and by understanding what we are doing is difficult.' Main then repeats: 'it's difficult, you can't fail to be impressed by this project.'[12]

Indeed, because the tunnelling took place out of public sight and was carried out with remarkably little impact on London, apart from at a few sites where there was increased road congestion – such as at Finsbury Circus and Tottenham Court Road – the scale of what has been achieved may surprise Londoners when they finally get to ride through the tunnels. Every time I went down in the tunnels I was struck by three things: the sheer enormity of the task, the size of the tunnels – and the quiet efficiency of the workforce. Owing to the emphasis on health and safety, there was very little mess down there compared with many building sites I have visited. The implementation of such a wide range of safely measures undoubtedly added significantly to the cost of the scheme, but it is impossible to argue that corners should have been cut.

In several places, where the proximity of existing tunnels created a greater risk of accidents, particularly close monitoring had to be undertaken. At Liverpool Street station a stretch of the old Post Office Underground system,* now sadly defunct, runs just 3 metres above the new tunnel. A special network of sensors, using optical fibre that can detect movements as small as one hundredth of a millimetre, was installed to monitor

---

* The London Post Office Railway, which operated between 1927 and 2003, carried mail along narrow tunnels between Paddington in the west to Liverpool Street in the east, stopping off at other mainline stations and sorting offices along the way. A stretch of the network was opened to the public in September 2017 as part of the new Postal Museum near Mount Pleasant in central London.

whether the old tunnel was deforming, but in the event the work passed off smoothly.

The most difficult section, which featured in the highly popular BBC TV series on Crossrail, *The Fifteen Billion Pound Railway* (2014 and 2017), was at Tottenham Court Road. Here the tunnel boring machines passed just 85 cm above the Northern Line tunnel and a mere 35 cm below the piles supporting the escalators. It was, as the programme called it, 'urban heart surgery' or like threading the eye of a needle. The man at the controls of the tunnel boring machine was Ed Batty who revealed to the cameras that he had only done this job for just over a year and that 'the first six months were a learning curve'. But, he promised, 'now I know what the crack is'. He was right. Using the laser equipment that ensured the boring machines stayed on course, the machine safely negotiated its way through the labyrinth of existing underground tunnels and utilities beneath Tottenham Court Road. So confident were Crossrail that there would be no risk to the public that Underground services continued to run through the station as the tunnelling took place, although a contingency plan to close the station had been prepared because of fears that even a minor fall of debris would have caused passengers on the platform below to panic. In the event, however, passengers were blithely unaware of the momentous event taking place a few metres above them.

While the work of the TBMs was subject to innumerable delays – and the process was never easy, as Main explained so well – there were no major mishaps and, most importantly, no serious injuries. That, unfortunately, was not the case with the tunnels excavated and stabilized using the spray concrete lining technique covered in the next chapter (see p. 212).

Water is a constant challenge faced by any tunnelling project. The worst problem encountered by Crossrail was at the Stepney Green works, where the lower depth of the tunnels at both eastbound and westbound junctions meant they penetrated deep into the Lambeth Group stratum. The size of the junctions contributed to the difficulty as they were each 17 metres wide and 50 metres long. At first, it was hoped that the creation of a large number of 'dewatering' wells to extract water would prove sufficient, but because of the permeability of the soil, this proved ineffective. Instead, a complex process known as 'in-tunnel depressurization' was installed to limit the amount of water affecting the site during the works.

Crossrail had decided to route the section of the line in the Royal Docks, which is further east and both bigger and deeper than the sections on the Isle of Dogs, through the Connaught Tunnel built in 1878 for the rather confusingly named Eastern Counties & Thames Junction Railway. It was one section of tunnelling which did not need boring since the 600-metre-long tunnel, was already built but was in desperate need of repair. The tunnel, which runs under the navigable passageway between the Victoria and Royal Albert Docks, had been used by that railway and its various successors until 2006 when the North London line section between Stratford and North Woolwich was closed to make way for part of the Docklands Light Railway and had been disused since then. Not surprisingly, the tunnel was in a poor condition due to its age and neglect and it had also suffered damage in the 1930s, when ships with a deeper draught were starting to be used, many of which scraped the exposed top of the tunnel. As part of work to deepen the docks carried out in 1935, the central section of the tunnel had been narrowed, brickwork removed and steel segment arches installed to reduce the tunnel's roof thickness and to strengthen and stabilize the tunnel.

This rather jerry-built adaptation was to cause a major headache for the Crossrail team. Originally, they had planned to remove the steel, backfill the sides with concrete foam and then run a TBM through to excavate it to the right size. Linda Miller, the engineer in charge of the project, says that surveys revealed that this could have resulted in disaster: 'To our alarm, we realised that using a tunnel boring machine could bring about a complete collapse of the tunnel and bring the dock water flooding in. This revelation about the weakness of the tunnel where it runs beneath the dock meant that our "Plan A" was no longer feasible. We had to go back to the drawing board to find a different solution to the problem.'[13] The answer, which was not cheap as the project eventually cost £50m, was to drain the area between the Victoria and Royal Albert Docks using a cofferdam (a temporary subterranean enclosure that is pumped dry to allow construction work to take place below the waterline) the size of a football pitch to dry out the roof of the tunnel.

However, the project faced significant obstacles and hazards. London's docks had borne the brunt of the Luftwaffe's attentions during the Blitz, and this was accordingly the section of the Crossrail works where it was most likely that unexploded bombs from the Second World War would be encountered. A team of specialists was engaged who used armoured vehicles with magnetic equipment to investigate the ground around the tunnel, but no bombs were found. Another challenge was the proximity of London City Airport, which at times limited the type of work that could be undertaken as high cranes would pose a potential hazard to low-flying aircraft. And there were also time pressures: work on the tunnel, which started in January 2013, had to be completed before September of that year, as by then ships would need to use the waterway for a big exhibition at the nearby ExCel centre. In the event the project was successful and the dock refilled in time, but it was a close shave.

Disposing of the spoil from all the Crossrail tunnelling and associated work was a massive undertaking. Over the lifetime of the project, Crossrail excavated more than eight million tonnes of material (of which six million were carved out by the TBMs) and its disposal was always going to be a complicated and expensive headache. In terms of economics, the easiest disposal method was to remove it all by truck, but that was considered to be environmentally unacceptable and would have caused enormous congestion – as well as a safety risk – on London's roads. As we shall see, lorry movements – even on a quite limited scale – would be the cause of several fatalities. Had *all* the spoil being trucked, the number of lorry movements would have been at least a third greater for the project as a whole and the risk of further fatalities concomitantly increased. Cleverly, by finding uses for all the spoil, no landfill tax had to be paid. Terry Morgan comments: 'It was recognized that one of our trucks killing a cyclist or pedestrian was one of the main risk areas. So we minimized truck movements as much as we could.'[14]

Two-thirds of what, in their more relaxed moments, the Cross-rail managers refer to as the 'muck', was used to build up an extension to Wallasea Island in the River Crouch in Essex to create a wetland bird reserve. The plan, therefore, was to remove the spoil on conveyors through the tunnel portals and convey it by rail and ship to Wallasea Island and to other sites where the

Tottenham Court Road station.

Paddington station.

Bond Street station.

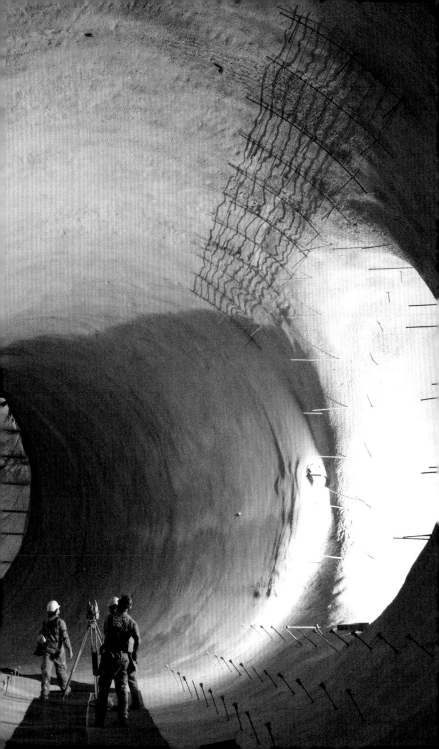

▲▲ Wallasea Island in Essex which was extended out of the spoil from the tunnels.

▲ Canary Wharf Crossrail station.

◀ Bond Street showing the platform edge doors.

The escalators at Whitechapel.

Elizabeth line
Westbound platform B ↑
← Heathrow
Maidenhead
Reading

Farringdon station, a new hub for the railway
network linking Crossrail and Thameslink.

The Crossrail train built by Bombardier.

The interior of a train showing all the carriages are linked by a passageway.

spoil would be useful. In the west, this worked well: the spoil was taken out to Royal Oak, just west of Paddington, and moved via rail and then boat down the Thames and north up the Essex coast to Wallasea Island. In the east, however, the plan to do the same hit a snag of the sort that it is difficult to predict. The spoil in the west was largely composed of clay and therefore relatively dry; the muck in the east contained much more water and, when it was loaded onto barges, it swilled about too much and risked causing the vessels to capsize. The Crossrail team attempted to drain the water away but that proved impossible, and, in the end, much of the muck had to be trucked. Ultimately about 20 per cent of all the spoil was carried by road.

It was not only Wallasea Island that benefited from Crossrail's mucky offerings. Indeed, because of the problems of transferring them by ship, other sites, nearer to Limmo, where the spoil from the eastern end of the tunnels was brought to the surface, had to be found. The dozen or so other sites which took spoil from Crossrail included a golf club in Ingrebourne, quarries in East Tilbury and Fairlop, a farm (which used it to improve grazing pasture for livestock) and numerous landfill sites in need of restoration.

Projects involving digging underground are required by law to undertake archaeological work if it is likely that historic sites will be disturbed. Crossrail was not only Europe's biggest construction scheme, but it was tunnelling under a long-inhabited site and consequently required an archaeological operation on a grand scale. Crossrail was the biggest dig ever undertaken in the capital and, to ensure nothing was lost, four contracts were established encompassing more than 100 archaeologists on 20 sites. Tens of thousands of objects and fragments stretching from prehistoric times to the early twentieth century were uncovered by these archaeologists and by the occasional observant worker. Gillian Tindall nicely summarizes the findings made at one site, the particularly fertile area underneath Liverpool Street station in the heart of the City:

> Not far off have come to light a Roman dagger, with its wood or bone handle rotted away, a sharp brooch-pin for holding a robe or a cloak in place, several Roman coins, Roman hair-pins and gaming counters, the remains of iron 'hipposandals' that preceded horseshoes, a medieval key... two Roman burial urns complete with ashes have also appeared, one falling pat into the hands not of an archaeologist but of a workman engaged upon the roof of a new sewer.[15]

As well as artefacts such as these, there were discoveries of a botanical nature, including the seeds of water plants that had grown alongside the long-buried River Walbrook.*

This is just a taster. The discoveries made by the archaeologists or, in some cases, churned up by the machines, were enough to fill a museum and, indeed, in 2017 there was an exhibition devoted to the *Archaeology of Crossrail* at the Museum of London Docklands. One of the most intriguing exhibits was a bowling ball found in what was once the moat of a manor

---

* A short and now subterranean river that rose near Finsbury and ran into the Thames just west of what is now Cannon Street station. The Walbrook flowed through the centre of the Roman settlement of Londinium. It gives its name to a ward of the City of London.

house in Stepney Green dating from Tudor times (when Henry VIII banned commoners from bowling on the grounds that it was the preserve of the aristocracy). Perhaps the most amusing item was a piece of Victorian porcelain, the remnants of a chamber pot with an amusing engraving on the inside of the bowl of a rather disconcerted fellow saying, 'Oh what I see I will not tell'. Medieval ice skates made from animal bones were found in the City, remains of flint tools from what appeared to be a toolmaking site of the Mesolithic period (about 10,000 to 5,000 BC) in North Woolwich – and no fewer than 13,000 pickle and jam jars under the old Crosse & Blackwell premises in Tottenham Court Road.

And then there were skeletons. Lots of them. The Crossrail works disturbed a burial site used for around 170 years, from the sixteenth to the eighteenth centuries, by the Bethlem Hospital (known colloquially as Bedlam). The site was thought to contain 20,000 bodies, many of them not those of residents of the hospital but of ordinary citizens whose parishes had run out of space for burial. Some 3,700 skeletons were uncovered by the excavations in 2015. In a four-week operation, carried out by a team of sixty archaeologists, the skeletons were brought to the Museum of London to be examined before being reburied at a site in Willow Cemetery on Canvey Island since no room could be found for them anywhere in the capital (thereby becoming the first Londoners to be transported by Crossrail). 'The only grave is Essex' ran the headline of a piece in the *Guardian* describing the skeletons' transfer to their new home.

The skeletons performed one useful service at least. No traces of live bacteria were found in any of these human remains, thereby proving Lord James's fears about the Crossrail excavations causing an epidemic of anthrax or bubonic plague to be fanciful. However, thanks to their discovery, a key scientific advance was made. DNA testing on teeth found in the Bedlam cemetery enabled scientists to confirm, for the first time, the identity of the bacterium which caused London's plague epidemic of 1665–6. It was *Yersinia 1* which confirmed that the individuals afflicted had indeed died of bubonic plague.

In west London, as predicted, few signs of old human habitation were discovered. However, parts of a 68,000-year-old antler and several fragments of bison bone were uncovered and identified, some with traces of gnawing, possibly by wolves. Other findings included a piece of fifty-five-million-year-old amber (fossilized tree resin), discovered at Canary Wharf, and parts of a woolly mammoth jawbone.

In the event, there was no major incident with the TBMs in relation to unknown objects. Three wells were encountered on station sites – which, being far bigger, had the greatest risks – and a couple of borehole casings were also hit, but none of these caused serious problems. The one incident that did result in a delay was when some steel piles, which had been revealed in the desk survey but had been thought to be outside the tunnel profile, were discovered to be on the tunnel path. In the event, the TBM had to be stopped and a temporary 'adit' – a timber-framed access tunnel – ahead of the machine had to be built to enable the piles to be removed.

As for ordnance, thorough precautions were taken to avoid any potential disaster. There had been, at the outset, as we have seen, considerable concern about the risk of finding unexploded ordnance, since much of the Crossrail route passes through areas of the City and east London that were heavily targeted in

1940–41. Crossrail developed a new strategy for assessing the risks from unexploded bombs, because existing methods proved to be unsuitable. A desk exercise, assessing risk in relation to bomb densities, using the maps that are now available online of where every known bomb fell, proved to be useless since it merely confirmed that Crossrail passed through several districts where intensive bombing had taken place but failed to pinpoint where any remaining ordnance might be. The sinking of test bores was also unhelpful: not only was the process expensive and caused delays, but the existence of shallow obstructions, such as bricks and buried concrete, made the bores unreliable. The risk of encountering unexploded German bombs was, in any case, likely to be low, since the tunnels were being bored at a greater depth than any surviving ordnance which, at most, is found 7 metres below the surface and usually much less. Consequently, a new method developed by Crossrail entailed a risk assessment based on examination of wartime records and consideration of geological conditions. This is now part of the 'learning legacy' which Crossrail has bequeathed. There was specially detailed examination of the Connaught Tunnel, the greatest area of risk.

The end of the tunnel boring phase came on 26 May 2015, three days after Victoria successfully broke into Farringdon at 3 a.m. in front of a crowd of tunnelling workers. The machine excavated the remaining section of the station tunnel, thereby completing the linking of all of the Crossrail tunnels. The finishing of the work was marked a few days later by an official ceremony attended by Prime Minister David Cameron and the mayor of London, Boris Johnson.

There remained the small matter of getting rid of the machines, which, owing to their size and, particularly, the way they work, are very unwieldy. They cannot be reversed; nor are they capable of running through the tunnels they have created. This is because their rotating cutting heads, which laid the concrete rings to line the tunnels at the same time as boring them, are, at 7.1 metres in diameter, wider than the completed tunnels, which are just 6.2 metres.

All the options for disposing of the machines were expensive. There was no problem with the gantries and belts at the back, which could be dismantled and reused. However, at Farringdon it was realized there was no easy way to get rid of Phyllis and Ada. The two machines were therefore driven on tight curves off to the side at the ends of the platform tunnel and unceremoniously concreted in, along with the odd time capsule. By the time Victoria and Elizabeth arrived at Farringdon, there was more room to work in, since the station space had been excavated. Driving them off to the side and burying them behind concrete would have disrupted the fit-out of the station, while removing the parts and taking them back to the shaft at Limmo was thought to be too time-consuming. As Bill Tucker puts it, 'the quickest and least disruptive method for removing the TBMs was to dismantle and remove the main parts of the machines and then cut apart the TBM shells using oxyacetylene for removal'.[16]

The other four machines were also dismantled and lifted out of shafts at the end of the tunnel boring process. As the authors of the study of the performance of the TBMs wrote, 'neither arrangement was considered optimal as stripping machines backwards is time-consuming compared with the usual reception pit arrangement, where TBM components can be lifted out directly. Also, significant amounts of underground cutting and burning were required during the dismantling process, adding to health and safety hazards.'[17] Reality dictated that, in the confined context of central London, there was never going to be sufficient space to allow easy ways of removing equipment of such enormous size. The authors of the tunnel boring study are somewhat critical, suggesting that not enough time was allocated for the procedure: 'To reduce the health and safety risk, and programme risk of TBM assembly and disassembly in confined space, appropriate time and space should be secured for TBM reception arrangements.'[18] Although much of the equipment was reused, the process of dismantling it was extremely expensive.

As we have seen, the machines had excavated six million tonnes of earth, enough, in one of those comparisons beloved of Crossrail's PR people, to fill Wembley Stadium three times over. Now, however, with the main tunnels completed, the work on all the various passageways, connecting walkways and stations could be finished.

10.

Stations for the future

With the TBM contracts let, and the work set to start, the next stage was to award the central station contracts. In many cases, different contractors were taken on for the initial piling work and the main station works including the fit-out because these projects were so large. The first station contract to be granted was for Paddington in July 2011 and the rest were let over the following year. According to Terry Morgan, the chairman, 'we worked from outside in because in an ideal world you want to get your TBMs [tunnel boring machines] through before building the stations'.

The 'critical path' for the contracts established a series of deadlines to ensure that they were let on time. However, the Crossrail board was careful not to do too much at once, and contracts were let sequentially, with no more than a couple of big ones being put to the board at any meeting. These large contracts, of which there were around twenty, were worth some £200m–£300m per station, but these were all to prove gross underestimates. Innovation was the key: 'We divided up the contracts into packages,' says Terry Morgan,' but we could not be sure that this was the best value for money. So in the procurement process, after we compiled a list of the companies, we then got an agreement from them all that if there were a more innovative way of carrying out the work, the scope of individual packages could change.'[1]

There was more to the contracting process than just selecting the cheapest offer. 'It was rarely the case that we took the lowest bidder. The contracts were assessed 60 per cent on quality and 40 per cent on price, and later, when the supply chain was getting smart to that split, we changed it to 70–30. This was to stop companies bidding low, because it is in the DNA of contractors to bid low.' This inbuilt tendency on the part of contractors to bid unrealistically low was further reinforced by economic circumstances – the recession that followed the 2008 crisis was still having an effect. But Morgan and the Crossrail team were all too conscious that if a very low bid is accepted at the outset, it can result in time-consuming arguments about cost variations and details of terms over the entire duration of the contract.

Letting these contracts was, of necessity, a rigorous and demanding process. 'We had estimates for each of the stations, but we had to review the process constantly,' Terry Morgan observes. 'We had a very high-quality risk register – this was not just some spreadsheet that we ignored. Every contractor had to demonstrate that they used it.' As mentioned in Chapter 8, the emphasis was to ensure that Crossrail was one team with a single aim in focus.

By the time most of the contracts were being let, there had been a change at the top. Rob Holden, the chief executive since April 2009, announced in January 2011 that he was leaving and was replaced that summer by Andrew Wolstenholme. Holden claimed he was 'exploring new opportunities' and Crossrail issued a bland statement: 'We regret his decision to step down but respect it. He leaves the project in great shape and we are on firm foundations to maintain our momentum to deliver this critically important project for the UK, on time and within budget.'

Holden's background as an accountant proved useful during the difficult negotiations with the Treasury and the Department for Transport, but he later found himself isolated, as he fell out with the Crossrail board. Holden felt that the board – created as a buffer between the two sponsors (the Department for Transport and TfL) and the Crossrail team – was an unnecessary layer of bureaucracy of a kind that had not been required in his previous job overseeing the construction of the Channel Tunnel Rail Link (now HS1). As one insider suggests, Holden was also cavalier in his attitude towards the boroughs: 'He rather felt that the boroughs were irrelevant since we were driving this railway through whatever happened, but that was to underestimate the delicacy of the situation. In fact, we needed the boroughs as otherwise they can make difficulties over, say, road closures or getting permission to demolish buildings quickly.' Holden left with his reputation intact as he went on to chair the Submarine Delivery Agency, one of the few projects that is even more expensive than Crossrail.

Holden's successor, Andrew Wolstenholme, who stayed until March 2018, is a very different character. A civil engineer by trade and a former army captain who had seen service in Northern Ireland during the Troubles, he had worked on the construction of Heathrow Express, which involved the construction of lengthy tunnels into the airport and used much of the same technology as Crossrail. He was also the programme director for the construction of Heathrow Terminal 5. A major tunnel collapse during construction of the Heathrow Express which fortunately did not result in any loss of life made Wolstenholme well aware of the potential risks involved: 'It was important we got press coverage that was benign, especially at the beginning. If we had got bad press in the early days, it would have been very difficult for us.'[2] While accidents or tunnel collapses would have elicited the most damaging press response, Wolstenholme points to the fact that Crossrail managed to avoid other negative publicity, as 'the procurement process went through without a single legal challenge'.

Wolstenholme arrived at Crossrail in August 2011 just as stations were becoming the main focus and, as he puts it: 'There was great momentum and a fantastic technical team getting the procurement right.'[3] Each of the ten central stations posed particular problems and required individual approaches. Depending on their location, they were designed either as 'box' or 'mined' structures. The box system essentially means cut and cover: in other words, the contractors dig a huge hole, create a cavity which will house the station and use piles to provide stability for the structure. This is a relatively new concept for station building in the UK, though it is used extensively elsewhere in places where there is sufficient space.

The box method represents a significant shift in station design, since it creates stations with a feeling of space, and also facilitates the installation of the escalators, lifts and ancillary equipment. This method was feasible for the sites at Paddington, Canary Wharf and Woolwich as they were shallow enough and had sufficient land nearby to create an effective worksite.

However, the scale of excavation involved means that these box stations do not come cheap.

It was, however, not possible to employ the box method in the more cramped conditions of the central London stations. The original idea was that work would start on these stations, from above, before the arrival of the tunnel boring machines, and this was how the construction was carried out at the eastern stations of Liverpool Street, Whitechapel and Canary Wharf. But the alteration in the programming of Crossrail, mentioned in the previous chapter, resulted in a change of plan for the construction work at Bond Street, Tottenham Court Road and Farringdon stations. At the request of the contractor* who had won the contracts for both the bored tunnel and the station at Farringdon, the TBMs would go through before the station works were carried out. Once the boring machines had done their job and continued on their way the station contractors would break up the rings that had been laid for the tunnel and start to excavate the space required for the station from the bottom up.

As we saw in the previous chapter, TBMs are now so sophis-ticated that by and large they pose little threat to the buildings and other structures above them. However, at the station sites, which involve much more extensive excavation and the con-struction of a range of different types of tunnel, there was a greater risk of creating undue settlement, of finding unexpected ground conditions or, more generally speaking, of suffering a variety of mishaps. Mike Black, the head of geotechnics, des-cribes the measures considered for the reduction of these risks:

---

* A joint venture between BAM Nuttall, Ferrovial Agroman (UK) Ltd and Kier Construction.

'Firstly, could the ground settlement be minimised at source by reconfiguring the station tunnels or altering the construction sequence? Secondly, could the movement be reduced by ground treatment such as permeation grouting of unconsolidated material or compensation grouting? Thirdly, could the structure or utility at risk be strengthened to better accommodate the expected movement?'

The main method for creating these spaces at the stations was through the spray concrete lining technique, though in a few places older techniques such as using steel ring segments were needed. In the spray concrete lining technique, the passageways are constructed by excavating small sections of the ground and quickly applying cement from pipes. This thin concrete lining provides temporary support for new tunnel openings where the ground conditions are not stable enough for them to be left without support for any length of time. The concrete quickly fills in cracks and openings, and prevents water from entering the tunnel. This is, however, only a temporary measure. A permanent, load-bearing lining, of either concrete cast in situ or precast concrete rings, will be required. TBMs cannot be used to construct these passageways, many of which were far larger than the tunnel diameter, because they vary in shape and diameter, and are short in length. The concrete used on the Crossrail project usually included steel fibres to strengthen the initial lining, which is normally around 7.5 cm thick. This gives a rather soft, natural-looking effect that has resulted in some remarkable and very photogenic spaces.

The construction of the stations was a task even more complex than the work of the boring the tunnels. These stations will be the public face of the Elizabeth Line – visible statements of a project that has been largely hidden from public view for the decade in which construction has being taking place. During the development of the scheme to build the Jubilee Line Extension in the 1990s, the previous major project in London, there was much debate about the kind of stations that should be built. Because the stations were the most costly element of the project, there was a good deal of pressure to build them on the cheap, using the least amount of space necessary. Every cubic metre dug underground comes at a fixed price, which is considerable, and therefore once the volume is assessed the gross approximate cost of the new station is simple to calculate. The government, which was footing the bill for the Jubilee Line Extension, recommended keeping costs down by building minimalist stations, but ultimately London Underground won out, arguing that the new stations, as visible expressions of modern London, should be a source of pride. London Underground went on to build memorable structures at all stations on the Extension, which have greatly enhanced the public realm and been widely feted.

And so it will be with Crossrail. In spades. From the outset, the notion was to create a coherent design strategy to ensure that Elizabeth Line stations would be instantly recognizable. This has been applied both to the ten large new stations beneath the city streets in central London and to the conversion of the existing surface stations. Even for the 1990s version of the Crossrail scheme, clear design standards had been adopted. The watchword has been to 'think big', and to design – and build – for growth. By the end of the first decade of the 2000s, with London booming and seemingly destined for ever faster growth, 'the station locations remained fairly constant [i.e. consistent with those of the first scheme], but they were radically redesigned with larger, often multiple entrances and ticket halls, working in parallel with big upgrades to the connecting Tube and railway stations'.[4] There were two major differences, however, between the earlier scheme and the one that was actually built: first, the new emphasis on 'sustainability', a buzzword of the early twenty-first century; and secondly the need for full accessibility, particularly for people with disabilities. The 1990s scheme emphasized 'lightness, minimalism, clear orientation for passengers and tough durable materials...'. Given that those principles have been retained, and that technology has progressed exponentially since that time ('today's railway is considerably more advanced than the one they almost built'[5]), the new Crossrail stations are bound to impress Londoners and other users.

The concept behind Crossrail has always been ambitious. Its top managers talk constantly of a 'world-class modern railway' and they genuinely mean it. Crossrail will be used by the grandchildren of people who have not yet been born, and its creators are very conscious of that. That is why there has been little skimping on detail and a strong desire to make each of the new stations a statement of intent. Much effort, too, has been expended on enhancing the existing surface stations which Crossrail will serve, both in improving accessibility and in implementing the design standards to make them visible manifestations of the Elizabeth Line. The aim has been to give a coherence to the whole line, following the lead of London Overground, which TfL rebranded with a distinctive orange livery and roundel in 2006. Unlike London Overground, which comprises a rather disparate and complex hotchpotch of lines, the Elizabeth Line is simpler, and therefore it will be easier to build up an identity for it. London's iconic transport map has run out of basic colours and the one chosen for Crossrail, a dark purple, is distinguished from the Metropolitan Line magenta only by its stronger tone. Nevertheless, it will be everywhere, from the benches and hand-rails of the steps to the trains themselves – and, of course, the relevant roundel.

Crossrail's determination to achieve the highest standards of design is well illustrated by the fact that they went to the trouble of creating a dummy station in a huge shed near Leighton Buzzard in Bedfordshire. According to Hugh Pearman, it had 'two platforms and a connecting passageway for passengers, a live overhead wire for the (mysteriously absent) trains, an esca-lator, a variety of signs, seats, lighting, balustrades, PA systems pretty much everything you'd expect to find down at platform level when the Elizabeth Line opens'.[6] This attention to detail may seem wasteful in an age of austerity but since Crossrail is a £15bn scheme, spending a few hundred thousand on getting the look and feel of the stations right is small beer.

The sprayed concrete technique has the advantage over older methods of creating a 'soft' environment, with no hard edges and with ceilings merging seamlessly into walls and floors. The generous size of the passageways, designed with growth in mind, will offer a striking contrast to the cramped trains and corridors of the Tube lines. The test station in Bedfordshire was used to assess an array of different features, including passageway lin-ings, floor tiles, lighting and even the escalators: 'Nothing was left to chance. Even the invisible bolts that hold together the complex multi-purpose wall separating platform from track were specially designed and tested by Imperial College to make sure they cannot work loose.'[7]

While there is a clear set of design standards, every station will look different. This is not just because of its location and its surroundings, but because Crossrail has given each one a particular identity to match the historic character of the locality it serves. Thus the design of Farringdon reflects the fact that it serves London's jewellery and diamond quarter in Hatton Garden; while Liverpool Street has a City theme whose designs reflect the traditional office wear of the men working there: 'The grooved, angled ceilings could be seen to resemble the pinstripes, often seen in the suits of City workers,' says the Crossrail website, though one suspects that not many of the millions passing through will spot that. But it makes for a nice surface.

Not all of the stations are masterpieces, but they are likely to instil a deep sense of pride for many Londoners. There will be, as there has been with the Jubilee Line Extension, much debate about which one is the best. Paddington, which is only just below the surface as it is near the western portal, will certainly be a contender. It has the advantage of being the only one of the new central London stations to enjoy daylight shining through to the concourse and the station platforms.

The choice of location for the Crossrail Paddington station was made for the 1990s version of the project when it was decided to place it on the southwestern side of the station. This was home to the prosaically named Departures Road and the main taxi rank, which plunged sharply down from the street at Eastbourne Terrace to the station level, Brunel having built Paddington in a dip. Moving the taxi rank to the northeastern side of the station was the first task for the Crossrail Paddington project when work started in 2011. That involved creating a spacious new concourse next to the canal basin for a new

Hammersmith & City Line station, few of whose users would have realized they were the lucky beneficiaries of a side effect of the Crossrail project, as the line is separate from the other Underground services at Paddington.

Only after the new Hammersmith & City Line concourse had been built could work start on creating the huge box that would house the Crossrail station on the station's southwestern side. Starting at a level above the main station concourse gave Crossrail an extra floor to use, which allowed for the transparent glass roof as well as permitting level access to the station from the concourse. The station was excavated from the top in stages and the team began by digging out the 260-metre-long box that would house the entire station. In April 2012 work commenced on construction of a 40-metre-deep reinforced diaphragm concrete wall* around the perimeter of the new station. Over the next eleven months, wall panels were installed around a massive box of 265 metres long by 22.5 metres wide. As ever with Crossrail, the statistics are startling: construction of the completed wall panels required over 64,000 tonnes of concrete and 5,000 tonnes of steel, and to set them in place more than 53,000 tonnes of London clay were removed. In addition to the 165 diaphragm wall panels, fifty-one plunge columns were installed to support the new station.

Crossrail was exceptionally fortunate in the choice of this excavation site because the huge space next to the station under Eastbourne Terrace did not contain any of the four Underground lines or the Post Office Railway which serve Paddington, leaving ample room for the Crossrail station. There was no major sewer

---

* A diaphragm wall is a reinforced concrete wall which is set in situ.

either and the various existing utilities could be moved, albeit, as ever, at great expense. Nor did the removal of spoil present significant logistical difficulties, since it only needed to be taken to Royal Oak station, about a mile away, for rail transfer. This was carried out in an operation that involved ninety movements per night, which consequently could be handled by a small number of trucks, and the short distance ensured there were not too many neighbours complaining about sleepless nights to appease.

For all the natural advantages of the site, building an enormous box station at Paddington remained a more expensive option than the conventional alternative of simply excavating whatever holes were needed. But, as a 2015 article on the *London Reconnections* website puts it, what tipped the debate in favour of building the box was, strangely,

Fans. Paddington station is a major site for managing airflow – especially in the event of emergency. Even in normal conditions one has to bear in mind that the trains will have air-conditioning, so the tunnels are liable to get very hot in summer unless something is done to get rid of the heat. The enormous fans that will be located there will work in tandem with similar enormous fans at Farringdon to control the air in the tunnels. Associated with the fans are separate emergency access shafts that will be guaranteed to have air as fresh as the streets outside in the event of fire – very useful for emergency safe evacuation or as an access route for firefighters.[8]

When the first layer of excavation was completed, the roof slab was used as a strut since the next layer down was excavated, by which time the TBMs had been through the station. By the summer of 2014 the station team had access to the tunnel section, and were then able to excavate down to the final level, around 25 metres below ground level. James Wainwright, the construction manager, explained to me the next part of the process on my site visit: 'When the platform level was built, it was handed over to the services – mechanical, electrical, plumbing and heating – and all the architectural elements. Logistically it has been difficult because it is a very narrow site and we had a commitment to reopen the road after only two years, which gave us far less room to work in.' As with every aspect of this project, there is always a number that pops out – at Paddington, just for the station itself, 300 kilometres of cable were used for the various services and communications equipment.

A major task during construction of the station was the removal of some of the 'plunge columns' used as temporary supports, so as to allow Eastbourne Terrace to reopen to buses in February 2014. A few of them, which were needed permanently to carry the weight of the structure, were encased in concrete to create elliptical columns that were then decorated with bronze panels from base to head height. Wainwright believes this was a very difficult process, 'as they were cast in situ with self-compacting concrete'.[9] He explained proudly how, although the pillars were cast in sections, none of the joins can be seen, even from close up.

The platform level at Paddington is atypical of Crossrail stations because the platforms are an 'island' between the two tracks, whereas at the other new underground stations the platforms are separate. Paddington's Crossrail platform, like all the below-surface platforms on the new line, will have doors at its edge that will only open in tandem with those on the train. Although nearly all new metros currently being built across the world incorporate such platform-edge doors, in London only the Jubilee Line Extension stations have this safety feature, another illustration of just how different the Elizabeth Line will be from most existing Underground services.

As there are three doors on each carriage of the nine-coach trains, every platform will have twenty-seven of these doors and every one has to work every time. Unlike those on the Jubilee, the platform barriers will reach ceiling height, which means that the hot air of the tunnels pushed through by the trains will not reach the platforms, so commuters will not experience the characteristic wind rush they enjoy in the Tube system before the arrival of the train.

As well as its 'elliptical column heads', Paddington can also lay claim to having the most bizarre artwork. The Crossrail board had wanted an art programme to go with the project but TfL and the Department for Transport, anxious to avoid accusations of wasting taxpayers' money, vetoed the spending of any public funds on it. Michael Cassidy, a long-time member of the City of London Corporation, took on the role of leading the art programme with the backing of chairman Terry Morgan and started trying to raise funds to pay for it. Cassidy realized that he would also need support from the Corporation and negotiated a canny deal. As mentioned in Chapter 7, the City had committed itself to passing the hat round to raise £100m on top of its original promise of paying £250m of the cost of Crossrail. However, because of the recession, it was unlikely that any business would be in the mood to contribute extra funds voluntarily. Cassidy therefore asked the City Corporation if it would provide £3.5m for the art programme if he could persuade the government to drop the moral obligation on the voluntary fund. Ministers realized that the extra £100m was never going to materialize and therefore agreed not to pursue the City for it, which meant that the future of the art programme was assured. In other words, the City was let off its £100m voluntary contribution if it paid £3.5m towards the art.

Cassidy sought contributions to specific stations and each one was paired with a gallery to help with the selection. George Iacobescu, the boss of Canary Wharf Ltd, immediately agreed to pay towards its station and others came forward quickly such as British Land at Liverpool Street, Heathrow at Paddington and Selfridges at Bond Street. Whitechapel proved the most difficult, not surprisingly because of its unfashionable location but eventually the money was found.

As a result, seven of the new Elizabeth Line stations will feature works of public art. Large collages produced by Chantal Joffe will decorate the platforms at Whitechapel, and contrasting works by two Turner Prize winners, Douglas Gordon (1996) and Richard Wright (2009), will embellish the ticket halls at either end of Tottenham Court Road station. At Farringdon, Goldman Sachs funded Simon Periton's glazed images of diamonds.

It is Paddington, however, that will have the most innovative installation, though many passengers may not notice it as they pass through. The American artist Spencer Finch has used the roof above the concourse to create a 'cloudscape', with different types of cloud printed through a mosaic-style technique into the glass panels. Terry Morgan, who was heavily involved in the art programme and who showed me round the station just after the work was installed, sounded a little doubtful when he promised 'it will be one of the iconic images of the railway'. He explained: 'It is a cloud index, representing clouds across the whole of the UK. During the day it looks just like the skyline. Each panel is unique, and you will see different types of cloud, nimbus, cirrus, stratus etc.' Finch's installation will be easy to clean, since it is made of triple-laminated glass that can be walked on or even – Morgan stressed – hold up scaffolding. Paddington's cloudscape looks certain to be one of the most talked-about features of the new line when Crossrail opens. And Londoners will have a new game to play of trying to distinguish fake clouds from real ones.

The most difficult new station to build was probably White-chapel, which had to retain the Victorian frontage of the old Metropolitan Line station and accommodate the existing London Overground railway as well as the new Crossrail tunnel. As Hugh Pearman explained: 'The north–south Overground in its brick cutting runs beneath the east–west Underground while the new tunnels sweep through slightly to the north at a deeper level. How to reconcile all this and make pedestrian links in the hinterland to the north with its school, sports centre and housing?'[10] Add to this the fact that Whitechapel will become an interchange station with only 20 per cent of its users actu-ally entering or leaving the system there, and this means the passageways have to be large enough to accommodate what will become a major flow at peak times. The demand at these inter-changes can be unpredictable as people often adapt their jour-neys depending on convenience and therefore it is best to build in considerable contingency when designing them. For instance, Canada Water, a new Underground station built for the Jubilee Line Extension, has also, since 2010, been on the Overground, and consequently its use has far exceeded predictions because people have found it to be a handy interchange between the two railways.

The answer at Whitechapel was to build a station concourse parallel to the Overground cutting at a new mezzanine level – no easy task because the site was already so crowded both at ground level and below. An additional challenge – on top of its two other railways and Victorian frontage – arose from White-chapel station's proximity to the northeast storm relief sewer, made famous by its tendency to produce huge 'fatbergs' built up from the drainage of the numerous local restaurants in the vicinity. The sewer was in a poor state of repair and, because of the disturbance that Crossrail caused to the area surrounding it, 'Thames Water required Crossrail to carry out some modifica-tions to strengthen the sewer… that took longer and [were] more difficult than expected'.[11] The new station at Whitechapel was plagued by other technical hitches: problems with its diaphragm walls – which are made by a technique that uses bentonite mud to provide temporary support for the excavation – took a year to resolve, delaying the programme for the site. The 3D diagram of Whitechapel station shows just how difficult it can be to build a new station around all the existing structures.

As was the case elsewhere, work on the new station also embraced local improvements to the public realm. At Whitechapel they spread out in all four directions. A public footpath will run right through the station, providing a new connection for people living north of the station with the High Street to the south, and a landscaped public square will replace what used to be a car park. To the west of the station, a bridge over the Underground tracks will be made vehicle-free and a widened pavement on White-chapel Road will link across to the Royal London Hospital.

Whitechapel was not the only station where the tunnelling caused problems given that, as explained earlier, the passage-ways were bigger and more complicated than those used by the trains. Finsbury Circus, which lies between the two entrances of Liverpool Street station, was the site where the ground conditions were most challenging and tackling the problem resulted in a year's delay in the contract for Liverpool Street Crossrail station. As chief executive Andrew Wolstenholme put it, this was where they came across the worst of the 'unknown unknowns': 'There was an escalator which was designed to come out of Moorgate westwards and the geology was just unusual. We had no records of any foundations of the building we were digging under, there was the Northern Line close by and the Thames Sewer but the ground was not responding in the way it should do. You can't take risks, you have to work with the engineering.'[12]

There was greater settlement than predicted and, in an attempt to stabilize the ground, the Crossrail team resorted to the standard method of injecting grout through pipes into the soil. But nothing happened. The ground did not stabilize, however, and more and more grout was injected into the ground. By January 2016, the *Ground Engineering* website reported, 'After almost five years on site, over 7.8M litres of grout has been injected so far from 12,500 *tubes à manchette* [tams] to support the buildings of Finsbury Circus and shallow excavation work for Crossrail's Liverpool Street station.' The 'tams' are up to 75 metres long, though their average length is just 50 metres (say a bit more than two cricket pitches long) and more than five years after being installed in 2012 were still in use. Yet, this huge amount of grout just seemed to be disappearing into the morass of sand and gravel and was still not having the desired effect. The problem was that some of the rather elegant buildings around Finsbury Circus – whose occupants had once overlooked a bowling green in the centre of the circus, but since 2011 had had only the view of a big hole in the ground with lots of orange-clad workers to entertain them – dated from the nineteenth century, and the nature of their foundations was unknown.

This part of the project was already a year behind schedule and Wolstenholme and his team were beginning to wonder whether they should implement Plan B: 'The last chance you get before deciding to simply find somewhere else to tunnel is to freeze the ground, which is difficult, lengthy and costly, and when it thaws you are not sure what you get. We were within a month of doing that and if that fails, you have to look for somewhere else to dig your tunnel. Then, at last, the ground stabilised and we could excavate the tunnel for the escalator.'[13] There were problems, too, at the Liverpool Street end of the station, where a further 3,500 'tams' were used to inject grout. But the challenge here was less acute than it had been at Finsbury Circus, as only three buildings needed grout to stabilize their foundations.

Another unusual feature of the work at Liverpool Street was that the escalator shafts were excavated from the bottom up. This was the first time that uphill excavators, commonly found in the mining industry, had been used in the UK. The uphill excavator dug the shafts using a pilot tunnel to control settlement, and then enlarged them to the full size and, as the newly dug shafts approached the surface, grouting was used to prevent settlement.

The surface stations on the Elizabeth Line – even the relatively lightly used ones in suburban London – have also been much improved. All their furniture, including seats and handrails, has been given the deep purple makeover. For example, West Drayton station, on the western outskirts of London, had particularly poor access, passengers having to negotiate narrow footways along Station Approach, a busy bus route. There were, furthermore, no amenities around the station. Crossrail is creating a new 'pocket park' next to the Grand Union Canal to the north, providing new access to the station from the waterfront, and Station Approach will benefit from the installation of a much better continuous footway on both sides.

At Maryland station, between Stratford and Forest Gate on the other side of London, there will be a new station entrance on Leytonstone Road and a wider pavement, together with the removal of a roundabout, will reduce the dominance of vehicles in the area. There will also be additional cycle parking, a taxi stand, and a pick-up and drop-off point. Changes of this type may appear modest, but these micro measures can make all the difference in terms of accessibility and of people's perception of the local area. Every station on the Crossrail route will benefit in some respect, and several will be entirely transformed.

Custom House, on the Abbey Wood branch, provides an illustration of one of the other techniques used to build the stations. It is on the surface, but was previously only a DLR station, and therefore a new station had to be built for Crossrail. The new structure was essentially built 'out of the box' with the concrete sections being prefabricated in Nottinghamshire and sent to the site in Docklands on a 'just in time' basis. While the simple structure of a steel-framed grid housing the concourse had to be positioned above the tracks because of the constricted nature of the area, the station needed to be big enough to cater for large crowds as it will serve the ExCel exhibition centre.

It is not just the spaces inside the stations that will see major changes. The Crossrail project has given TfL the opportunity to redesign and expand the spaces around the stations. Working with the local boroughs, this will result in '40 improved spaces outside stations', twenty-four new and a dozen improved forecourts or – expressed in one of those comparisons favoured by Crossrail – 'the equivalent of 19 Leicester Squares'. Or at least this is the aspiration. The cost of these improvements is £130m and around £100m had been raised at the time this book was going to press, with the money coming from Crossrail, TfL, local authorities and developer contributions. Crossrail remained confident that all these improvements would eventually be delivered.

It is difficult to know, at this stage, which of the new central stations will be the most spectacular or become Londoners' favourite. They will all, in their different ways, be contenders. There is not the space to do justice to them all here, but suffice to say that each one is larger than any existing London Underground station, apart from such recently modernized ones as King's Cross St Pancras and Victoria. Building just one station on this – unparalleled – scale in the centre of London would have been in itself a major achievement. Constructing ten of them simultaneously seemed, on the face of it, to be impossible and the sheer size of the task, did, it has to be said, tax the available capacity of the nation's construction industry. (I must add, as a historian of the London Underground,* that Leslie

---

* Covered in my book *The Subterranean Railway: How the London Underground Was Built and Changed the City Forever* (Atlantic Books, 2004).

Green's achievement of designing and supervising the construction of no fewer than fifty arts and crafts stations in central London in the space of a few years in the first decade of the twentieth century, most of which survive largely unnoticed by Londoners, is comparable, and, sadly, killed the poor man of exhaustion-induced tuberculosis at the age of thirty-three. But Green's stations were modest affairs in comparison to Crossrail's gargantuan caverns.)

As we have seen, just twenty people worked on each of the tunnel boring machines. The construction of the other tunnels and the stations, and the fitting-out of all the excavated spaces was, however, far more labour-intensive, some 10,000 workers being required at the peak of the construction work. While there were no serious incidents with the TBMs, one fatality occurred during the construction process, and it involved spray concrete lining. On 7 March 2014 René Tkáčik, a Slovakian, was spraying concrete onto excavated surfaces of a cross passage in the 24-metre-deep Fisher Street shaft in Holborn when a section weighing around a ton broke off and killed him. Tkáčik was an experienced shotcrete worker, but when the incident occurred he was working directly beneath the crown of a cross passageway that had just been sprayed. This was against the rules since a newly concreted area is considered an 'exclusion zone'; very few people are allowed in this exposed area precisely because of the fear of concrete or earth falling on them. At the subsequent inquest, the coroner suggested that Tkáčik's limited English may have resulted in a misunderstanding during briefings that warned workers not to go under the crown of recently sprayed shotcrete and criticized the contractor for not putting up a physical barrier preventing people from entering the area. The jury recorded a 'narrative verdict', noting that the definition and supervision of the exclusion zone was unclear. The contractor, Bam Ferrovial Kier, paid £350,000 to Tkáčik's family in compensation and was later prosecuted by the Health and Safety Executive over this incident and two others, involving respectively an accident with a tipper truck and a worker suffering severe injuries after being hit by a high-pressure mixture of water and concrete, both also on the Fisher Street site. Bam Ferrovial Kier, a partnership of three large companies, was fined just over £1m with costs of £42,000 in relation to the three incidents.

Strangely, Crossrail's worst safety record was outside its worksites and involved the lorries carrying material to and from them. The difficulty of obtaining access to Crossrail sites has been one of the contributors to the high cost of the project and this logistical operation is just as complex as organizing what happened below ground. There had been 770,795 lorry movements for the project by May 2018, and the very fact that such a precise number is known demonstrates that every movement is carefully monitored. Movements have to be tightly regulated because the worksites cannot accommodate any extra vehicles hanging around to drop off their loads; each arrival and departure had to be precisely scheduled.

Tottenham Court Road was a particularly constricted site. Work there started in 2009 when the existing Underground station was rebuilt and greatly extended, after which the new Crossrail station was constructed. For nine years this tightly packed corner of London, next to the Centre Point building, has had to accommodate a stream of lorries entering and leaving the site. Since the presence of even one lorry waiting to enter would cause traffic chaos, a holding station was set up on a spare section of road near Holborn station half a mile away, where they waited to be called through. As I wrote after a visit in 2012, the site was so constrained that 'even the dumping of a single extra skip requires planning and organisation'.[14]

This logistics planning was so complex that it merited an entire article of its own in one of the two special Crossrail supplements produced by *New Civil Engineer*. It also required a clear set of regulations. Again, the numbers are impressive: 'Over 800 hauliers delivering to approximately 30 main contractors across more than 50 central London sites, the largest construction project in Europe had to take the initiative and set a standard for all contracts to follow.'[15] Crossrail, rather reluctantly at first, decided that it had to be responsible for setting standards because it 'was always going to be held directly accountable by sponsors and stakeholders for any impacts outside the gate'.[16] That has certainly proved to be the case: Crossrail received considerable negative attention and criticism after a series of fatalities caused by its lorries.

This was new territory. Crossrail was one of the first projects to mandate that lorries making deliveries were contractually required to fit additional safety equipment and that its drivers undergo training to ensure the safety of cyclists and pedestrians. Crossrail lorries were required to travel on specified routes in central London designed to limit the number of left turns, a known hazard for cyclists. More than 10,000 drivers undertook these extra safety courses. As the scheme progressed, in 2012 Crossrail increased the number of safety requirements and training courses, and lorries failing to meet the requirements were supposed to be turned away from sites. On occasion, however, some lorries seem to have slipped through the net as, the authors of the *New Civil Engineer* article noted, 'these requirements were new to much of the industry, which, coupled with pressures to meet the programme, led to inconsistencies in the compliance checking process and a reluctant to report non-compliance in case it was perceived as poor performance. As a result, some non-compliant vehicles were allowed on site.'[17] When these failures came to light, Crossrail increased the pressure on the hauliers and managed to achieve higher rates of compliance. In spite of the cost and difficulties entailed in setting up such a scheme, several other big projects have taken an interest in adopting similar policies.

Despite the project's emphasis on safety, four people were killed by Crossrail contractors' lorries, three of whom were riding bicycles. With the death rate for cyclists in London hovering at around ten a year, this represented a very high proportion of casualties in the two-year period during which these accidents took place. First, in September 2013, Maria Karsa, aged twenty-one, was killed by a lorry in Aldgate. Because of roadworks, the lorry was not following an approved route. Just two months later, Brian Holt, aged sixty-two, was killed in the Mile End Road while wheeling his bike across a pelican crossing in front of the lorry, which was stuck in traffic. The third victim was Claire Hitier-Abadie, aged thirty-six, who in February 2015 was hit from behind at a junction in Victoria. The fourth fatality, Ted Wood, aged seventy-four, was a pedestrian who was killed in February 2014 by a Crossrail contractor's van at Aldgate. With the exception of the lorry that killed Maria Karsa, all three other vehicles were on Crossrail's approved routes and all four drivers had completed Crossrail's training course on cycle safety. Prosecutions were attempted in all but the first case, but only the driver in Victoria was convicted of careless driving, receiving a ban and a community service order.

With the tunnels completed, and work on the stations well underway, Crossrail now began to focus on what sort of railway it was creating.

11.

Trains and tunnels

One of the most far-sighted decisions made by Crossrail's top management was to appoint an operations director early in the process. When I interviewed the man who got the job, Howard Smith, in May 2018, he had then been in his post for five years, and I joked that since his appointment he must have sat about not doing very much since there were not yet any trains to operate. He laughed. In fact, Smith has been a busy man throughout this period and has ensured that the ultimate purpose of the project – the operation of trains rather than brilliant feats of engineering – has been a key focus of the project's senior management team.

After being appointed, Smith wasted little time before setting in place the processes for finding an operator for the railway and choosing the trains that would run on Crossrail. The initial call for bids from operators to run the system had been made before his arrival. The prestigious nature of the project meant that there was no shortage of interest. Four companies, all with experience of running trains in the UK – Arriva, National

Express, a joint venture involving Keolis and Go-Ahead, and the Hong Kong-based operator MTR (Mass Transit Railway) – were eventually shortlisted and bid for the operating contract. Unlike the franchises which are standard on the national rail network, the scheme to run the Crossrail trains was designed as a simple management contract, since all the revenue from the fares goes to TfL.

This was a similar arrangement to the London Overground, which had long been run by MTR (though it has subsequently lost the contract). There was therefore little surprise when MTR was announced as the winning bidder in July 2014. The contract involved granting an eight-year concession to operate Crossrail services, said by TfL to be worth £1.4bn, with the potential to extend it for another two years if the performance proved satisfactory. The Overground is very successful and Smith was determined to ensure that the lessons learned from MTR's running of that railway were applied to Crossrail. In particular, Smith is keen to ensure that MTR should know exactly what was being required of them, something which often does not happen with franchisees on the national rail network: 'We made sure that in drawing up the contract for the concession to operate the trains, our concept of what Crossrail was to be was made clear. It is all down to the detail, such as how many people are going to be on the platform, how the signalling system will work, where the control rooms will be, where the trains will be stabled, where the drivers' accommodation will be – all that dull, slightly techie second-order thinking without which you end up with a big mess.'[1]

The bidding process was not structured in the conventional way whereby the lowest bidder always wins. Instead, demonstrating the ability to deliver a high-quality service was crucial to winning the contract. Crossrail wanted a long-term partner to manage the preparation for the new service, to ensure its smooth introduction and to establish a clear modus operandi that will have a lasting impact. MTR also had the advantage of running a similar service to Crossrail in Stockholm. While there is a fixed fee for running the service, there are bonuses and penalties, and, interestingly, the latter will be imposed for delays even when they are not the operator's fault. That ensures there will be no games of the 'not our fault, guv, it was Network Rail' variety, which are widely played on the national network.

Smith explains: 'We wanted to incentivize them to be on the case for absolutely everything. They will also get paid on customer satisfaction and on ticketless travel, which has gone dramatically down on the service we already operate from around 5 or 6 per cent when we took it over to about one and a half per cent. That's by gating, by putting people on barriers and by full staffing, which of course we do.'[2] Jeremy Long, the chief executive of the European wing of MTR, realized that this way of operating required a culture change based on co-operation rather than confrontation, and created joint teams consisting of both Network Rail and MTR staff: 'Dedicated performance teams and customer experience teams worked closely together to analyse various measures such as station dwell time, in order to identify solutions and improve "right time" statistics. Such activities show that best practice is constantly being worked upon and improved through greater collaboration.'[3]

While this might sound a bit like PR gobbledegook, it actually addresses a fundamental problem created in the rail industry by the fragmentation necessitated by its privatized structure.

Failure to work in a collaborative way will undoubtedly affect Crossrail's performance and therefore reduce its positive impact.

In line with the policy of introducing Crossrail services gradually, rather than with one 'big bang', the contract involved the winning bidder taking over the suburban service between Liverpool Street and Shenfield in May 2015 and operating it under the London Rail banner, first with existing old trains and then, from 2017, gradually introducing the new ones. These were, however, restricted to just seven cars, rather than the nine that will be used when the full service comes into operation, because some of the platforms were too short to accommodate the full-length trains. The idea was to introduce Crossrail services in stages, first running between Paddington and Heathrow in May 2018, then through the tunnels between Paddington and Abbey Wood in December 2018, then Paddington–Shenfield in May 2019 and, finally, the full 115-kilometre (70-mile) service between Reading and Shenfield in December 2019.

However, the plan to run Crossrail's own trains through to Heathrow in May 2018 proved impossible because of difficulties over the signalling. In the Heathrow tunnel, a system known as European Train Control System (ETCS) Level 2 is used, which is different from the Communications-Based Train Control (CBTC) system in the main tunnels. The fitting of the new Crossrail trains with ETCS Level 2 proved more difficult than anticipated and therefore the existing trains used by the former Heathrow Connect service remained in service for several months. This was something of a warning of potential difficulties to come. As will be explained below, Crossrail uses three signalling systems and therefore the trains need to be fitted with three types of equipment. This is by no means unique in modern trains, but it adds to their complexity and their cost, and can reduce reliability.

The process for selecting the trains themselves was carried out in tandem with the bidding for the operator. The procurement of the trains was complex and politically challenging. The last big order for trains, intended for Thameslink, had gone to the German company Siemens rather than to Bombardier, then the only train manufacturer with a factory in the UK, which caused enormous controversy. There was, therefore, much pressure on the government to ensure that this time the order went to Bombardier. Otherwise the future of its Derby factory might be in jeopardy.

First, however, the question had to be settled of whether the trains should be acquired through a private finance initiative scheme, which was still the default for big public procurement programmes. A PFI deal would involve the financing of the trains by a private company, which would also take the risk on future maintenance. The fundamental justification for PFI, apart from the technicality that the cost is not included on the government's books, is that the private sector is more efficient than public organizations. The downside is that the cost of borrowing for a private company is always higher. The decision of whether to go for a PFI deal therefore rests on a judgement that the extra efficiencies as well as the innovation from private entrepreneurial flair more than make up for the extra cost of finance.

TfL, which was responsible for train procurement, examined the PFI strategy closely but found that there would be few advantages as the high cost of the finance would not be recouped in terms of savings. Smith says the idea that the PFI deal would have driven innovation and the development of a new type of train was misconceived: 'The train on offer would have been an updated version of the Thameslink model and therefore it would include little innovation. It was perfectly obvious that the bidders were going to put in what they put in for Thameslink, near as anything, so the extra cost of the finance would not have been worthwhile.'[4] And there was one crucial feature of the new trains that Smith and his team fought for 'like tigers':

> We insisted that the carriages should have three doors, not two, on each side. The key point is that this would greatly reduce dwell time, and if you are spending so much money on getting trains faster through central London, there is no point skimping on details. Most trains have only two doors per carriage [on each side] because many stations are on curves and therefore people using the middle one would have to negotiate a big step or alternatively the door would remain shut through using selective door operation now available on new trains. However, all the Crossrail stations both in the tunnels and on the surface are sufficiently straight to allow the middle doors to open and enable people to enter and exit safely, and therefore this problem does not arise.

In fact, the requirement of this distinctive arrangement of doors was another reason for not having a PFI arrangement. This would ensure that the trains were bespoke and would not be useable on other routes, and therefore their development costs could not be spread over other contracts. Consequently, the risk to the provider if, say, the trains were taken out of service halfway through their normal thirty-year life cycle, was greater as they could not have been deployed elsewhere and that risk would have to be included in the price of a PFI, pushing up the cost.

These factors all pointed towards a conventional funding deal. TfL decided to purchase the trains directly and entrusted the process to Crossrail, acting on behalf of the two sponsors, itself and the Department for Transport. Originally five firms were shortlisted – Hitachi, Alstom, Bombardier, CAF and Siemens – in June 2011, just before the announcement that Siemens had, surprisingly, won the Thameslink contract, a complex PFI deal, which sparked off major controversy because of the company's lack of a UK base. There were calls for the Thameslink process to be restarted and ministers found themselves having to defend a decision which they stressed was made in the interests of obtaining best value. Their argument that the deal would create a couple of thousand jobs in the British supply chain convinced few, and unions and politicians – of both major parties – tried to get the decision reversed. Bombardier warned that the decision might spell the end of its UK-based operation and threatened to make half its 3,000 staff redundant.

The Thameslink decision, nevertheless, survived this furore. But as a sop to its opponents, the government launched a 'review' of the Crossrail procurement process to ensure that 'the UK is making best use of the application of EU procurement rules, as well as the degree to which the government can set out require-ments and evaluation criteria with a sharper focus on the UK's strategic interest'.[5] It did not take much reading between the lines to suggest this was a way of trying to ensure the winning bidder created more added value in the UK economy rather than on the Continent. As the *Independent* commented at the time, 'In other words, the Government would dearly love to avoid another firestorm like that created when a £1.4bn contract to build rolling stock for Thameslink was awarded to Siemens, which plans to make the trains in Germany.'[6]

The review proved the last straw for Alstom, whose articulated* trains had not found favour for the Thameslink bid partly because Network Rail suggested they would damage the track more quickly than standard rolling stock. The French company pulled out in a strop, citing the continued delays in the process. The remaining four successfully prequalified to bid and, soon afterwards, TfL confirmed its decision not to go ahead with a PFI arrangement. Siemens then announced that it was withdrawing, stating that work on the Thameslink order would make it impossible to fulfil the Crossrail contract in time. In fact, Siemens seemed to have lost interest once the decision not to follow a PFI procurement process was made. There was a suspicion, too, that this was a tactical withdrawal in the light of the hostility which had greeted its successful Thameslink PFI bid.

Bombardier was announced as the winner in February 2014. It should be noted that, although Bombardier has a factory in Derby, it is a Canadian company whose main European office is in Germany. Train building is increasingly an assembly exercise: components are manufactured in many different places and then brought together at the plant. In the past, most parts were manufactured in the factory that made the trains and therefore

---

* Articulated trains have cars that are permanently attached to one another with fewer wheels and bogies, and therefore provide extra space for passengers and can result in savings both on initial costs and maintenance of the tracks.

an order created many more jobs in the winning bidders' plant than is the case today.

Howard Smith is adamant that the decision in favour of Bombardier, made in February 2014, 'genuinely wasn't a fix'. He says that while the headline price of the Bombardier trains was not very different from some of the other bids, the factor that swung it in Bombardier's favour was the whole life costs, since the contract includes maintenance of the trains for thirty-two years.

> It was a two-stage process that basically first assessed whether the train met the basic requirements and then it was largely on price but we included the depot in that as well and elements of whole-life costing such as the amount of electricity, track wear and, crucially, the maintenance over the thirty-year life of the trains. And it was the long-term maintenance costs that won it for Bombardier. All these factors could only ever have led to a single answer – and that was Bombardier. As it turned out, it was the right political answer and so the deal sailed through at TfL, but that was not the reason we chose Bombardier.[7]

The Bombardier bid differed from the others in one crucial way: rather than proposing trains consisting of ten carriages each 20 metres long, the suggestion was for nine carriages of 23 metres each, saving, therefore, on both the number of carriages and on track wear as there are fewer wheelsets. This means that the trains are 205 metres long (the two carriages at either end are a bit shorter), rather than the planned 200 metres. However, since the platforms of the line's central section were, in any case, designed to take twelve-carriage, 240-metre trains, there is no immediate issue. The platform lengths for the stations in the central tunnels vary, for local reasons, between 207 metres for Paddington westbound to 249 metres at Bond Street eastbound. If the trains are extended by two carriages, which will happen if the system becomes too overcrowded, work will be required at several stations to extend the platforms, but passive provision has been made to accommodate this. One additional carriage could be accommodated without much extra work at nearly all the below-surface Crossrail stations, though two of the three Heathrow stations have space only for 205-metre-long trains. On the surface, selective door opening – in other words, some doors not being used – will be needed at nine of the smaller stations such as Iver and Langley in the west and Maryland and Manor Park in the east because their platforms are shorter than the trains.

To house the trains, as part of the contract, a new £142m depot has been built at Old Oak Common, 6 kilometres out of Paddington on the Great Western Line. It has capacity to house forty-two of the seventy trains overnight and facilities to undertake heavy maintenance, such as replacing worn wheels, at the site. As the trains enter the depot, they pass over a remarkable machine called AVIS – automatic vehicle inspection system – which checks for a wide variety of faults automatically, such as door problems, wheel and pantograph wear and even graffiti, as the train goes slowly over it. Every train will go through the AVIS machine at least once every forty-eight hours to ensure no faults have developed. The other trains will be housed overnight in several places along the line – including Gidea Park and Abbey Wood – which are equipped with the necessary cleaning facilities. Housing the trains in different locations along the line ensures they can begin the service in the morning in a structured way, rather than all the trains having to start from a single point.

The Bombardier order was initially for sixty-five nine-coach trains, but this was later increased to seventy. One of the extra trains was required because of the decision to extend Crossrail to Reading rather than terminating at Maidenhead. Oddly, this led to a dispute over how the seventieth train would be paid for. The original £1.4bn cost of the trains was not included in the overall budget for Crossrail but is being funded by TfL, which has borrowed the money to pay for them. Although the Department for Transport approved the decision to go to Reading, it baulked at tossing in the extra cost of the train and argued that it ought to come out of the Crossrail budget, even though the other trains had not been included in it and by early 2018 it had become clear that the total spend on the project had exceeded the planned funding. While the matter may appear trivial, it is worth noting because it shows the constant pressure under which those who work on a project like Crossrail have to function. Eventually, extra funding to pay for the trains was made available by the government.

The trains, known as 345s, are the first of a new model, Aventra, developed by Bombardier to be lightweight and therefore causing less wear and tear on the track. They can carry a total of 1,500 passengers, of whom 450 are seated, and have a top speed of 90 mph, though they will have few opportunities to reach that. The 345, 'a Rolls-Royce of trains' as Smith describes it, is immensely complex since it requires three signalling systems – explained below – and is driver-only operated (known as DOO in the trade, and the source of enormous controversy with the unions, though not for Crossrail as the decision **not** to have guards was made very early in the process). The train is fitted with equipment that is able to capture pictures from the CCTV cameras on the platforms to allow the driver to have a clear view of the whole station – vital given their length. Very importantly for safety, the CCTV system allows the driver to keep an eye on the platform as the train is pulling away to ensure that no one has been trapped in the doors, though this will be impossible in the tunnel sections of the line because of the platform edge barriers. The trains will effectively be driven automatically through the central tunnel section. Getting the software right, to guarantee that the trains will stop precisely in the right location with the centre of the train doors aligning with those on the platform, requires accuracy down to a few centimetres. The key is to ensure this is done automatically, as it is a very difficult task for humans to perform within that level of accuracy as the trains will be slowing down from speeds of up to 60 mph as they enter the station.

As Crossrail found to its cost, being the launch customer for an exciting new train comes with certain disadvantages, since new train designs commonly suffer from a range of faults and do not, as hoped for, 'work out of the box'. This proved to be the case with the first trains used on the Shenfield line in 2017, which experienced a high rate of technical issues. But over time their performance has improved and it was, as of June 2018, far better than that of the old trains they replaced. It is not the hardware such as the motors or braking systems that is the problem but the software that is now the key component of any modern train. And that never works perfectly straightaway. One software engineer expert told me that there is likely to be one mistake, on average, in every thirty lines of code. In a safety-critical industry, such problems evidently have to be sorted out through thorough testing.

The most complex aspect of the trains is the equipment with which they have to be fitted in order to cope with three signalling systems. Even had there been no other difficulties, the complexity of the signalling system would have prevented the opening of the line in December 2018 and has been one of the major causes of the protracted delay.

So why *do* the trains have three systems? Here you need to take a deep breath; and those to whom the complexities of signalling systems are not of interest or who have an aversion to acronyms may even want to skip a few paragraphs...

The need for the three systems is a result of European requirements and the desire to create an interoperable system throughout Europe, which currently has a wide variety of signalling systems that make it difficult to run trains through frontiers. By European rules, all new railways must now be fitted with a system that is compatible with European Train Control System (ETCS) Level 2, which requires trains to be controlled by radio signals in the cab, obviating the need for lineside signals. The obvious solution would have been to fit ETCS Level 2 on the Crossrail trains, but to continue to use the Network Rail system – called Train Protection and Warning System, or TPWS – on the old surface sections west of Paddington and east of Liverpool Street.

However, the situation was further complicated by the fact that the signalling system, a type of Automatic Train Protection (ATP), in the Heathrow tunnels was different and while more sophisticated than TPWS, was not compliant with ETCS Level 2.

That left Crossrail with a difficult choice. Either the ATP system, which is reckoned to be redundant, would have to be fitted to all the trains, at great extra cost, or Crossrail would have to install an ETCS Level 2 system in the tunnels. While this is also costly, it is at least useful in the long term as, eventually, ETCS Level 2 will become the norm on Britain's railways.

It gets worse and more complicated. The obvious solution would have been to fit ETCS Level 2 in both the central and the Heathrow tunnels, but the Crossrail managers were concerned about progress in the development of the ETCS Level 2 technology. While it has been developed sufficiently to operate well on several railways in Europe, the level of sophistication required in the tunnels was far greater. The trains in the central tunnel had to be equipped with an ATO (Automatic Train Operation) system – which means the trains can be controlled by computers with no one in the cab actually driving the train – and have the capability for 'auto reversing'. This is required because around half the trains will terminate at Paddington and will need to be reversed rapidly. With this facility, drivers will be able to select 'auto reverse' and then leave the cab to walk through the carriages to the other end while the train travels on automatic through to nearby Westbourne Park where it can then reverse back to Paddington after crossing over to the eastbound track. By the time it reaches Paddington, the driver will be in the other cab ready to commence the train service back through central London.

The Crossrail team was worried that ETCS Level 2 was designed for mainline operation rather than a suburban network with a long central section in a tunnel. As well as this, ATO had to be integrated with the operation of the platform doors, and there was no system in the world with such complexity. Fitting ETCS Level 2 alongside these new features was seen as a major risk and Crossrail therefore applied to the European Commission for derogation (exemption) from the requirement to fit ETCS Level 2. The Crossrail team pointed out that while Thameslink was, at the time, having an ATO system fitted through the central section between St Pancras and London Bridge, it was not designed to have platform doors or the auto reversing capability.

The Crossrail team was haunted by the experience on the Jubilee Line Extension, where a similar attempt to jump a generation and introduce a revolutionary system initially proved unworkable, resulting in several years' delay before ATO could be introduced. The Crossrail team was determined to avoid similar problems. Although ETCS Level 2 is scheduled to be introduced throughout the Great Western lines, there is no target date and the Crossrail team, rightly, realized early on that it would not be ready in time for the original scheduled opening in 2018. There are, as yet, no scheduled plans for the ETCS system to be introduced in the east, so two systems – TPWS and ETCS – were always going to be necessary. After much consideration and hand-wringing, Crossrail ruled out the idea of deploying ETCS Level 2 in the central section as the system had not been sufficiently developed to accommodate the features they needed.

Instead, for the central section, a tried and tested generic system called Communications-Based Train Control (CBTC), first developed in the 1980s and subsequently much improved, was selected. The advantage of CBTC is that the location of the trains is calculated very accurately through a system of computers linked by radio using a series of lineside beacons and tags fitted to the trains. There are no external light signals and therefore the trains are able to operate more closely together because the computer continuously calculates the safe distance between them – technically this is known as 'moving block'. This allows a greater throughput which will reach, eventually, the expected peak-time traffic of twenty-four trains per hour in each direction.

Crossrail could not, however, obtain a derogation from the European Commission to **not** fit ETCS Level 2 for the Heathrow section. Chris Binns, Crossrail's chief engineer, explains: 'At Heathrow, the lower service frequency and the facilities do not require the fitting of ATO or auto reversing functions: the fleet currently serving Heathrow is conventionally driven, there are no platform screen doors and there is no need for automatic train operation or auto reversing to meet the service intervals.'[8] So there was no alternative to fitting ETCS Level 2 to the Heathrow tunnels.

The only other possible solution, to avoid the need for three systems, was to overlay CBTC on the surface sections of the railway, but this was considered far too risky. No railway in the world has such a long section using CBTC and the cost of fitting it would have been prohibitive. So that is the story of why Crossrail has three signalling systems – a situation which has, of course, resulted in a high level of extra cost and increases the risk of train failures.

This rather convoluted summary is shorthand for a much longer decision-making process that was fraught with complexity and which will have long-term implications for the railway. The mistakes made in the Jubilee Line Extensions not only cost tens, if not hundreds, of millions of pounds, but also damaged the reputation of London Underground, which then had the misfortune of suffering another expensive fiasco over the signalling contract on its sub-surface lines – the Circle, District and Metropolitan – in the early 2010s. This resulted in the need for retendering and – again – the loss of hundreds of millions of pounds. It was Crossrail's determination not to make the same mistakes that prompted the decision to use well-tried CBTC technology. But the need to operate three systems – TPWS, ETCS and CBTC – undoubtedly increases the likelihood that the early teething problems with ETCS, which resulted in the delay in the introduction of the 345s into the Heathrow tunnels, may be a prelude to future difficulties. The signalling, too, had to be bidirectional: in other words it allows eastbound trains to travel in the westbound tunnel and vice versa. This is essential not just for coping with emergencies, but also to enable parts of the tunnel to be closed early each night for maintenance, when the service will be so sparse that the last few trains might safely run on the opposite line, rather than their normal one.

Early on in the project, it was decided that the Elizabeth Line would be part of Transport for London – like the Underground – rather than owned by Network Rail, as is the case with Thameslink. This does create anomalies and there is no particular logic to the decision, except, perhaps, that Thameslink stretches further out into various parts of the South East and even the East Midlands and East Anglia – and has far more routes – than Crossrail. The question arose as to who would be responsible for the maintenance of the new sections of the line, particularly the tunnels. This could have been devolved to Network Rail but, instead, Crossrail is to have its own dedicated maintenance staff and, as with London Underground lines, a control centre, which will be located in Romford. It is from this new, purpose-built, three-storey centre just next to the tracks and completed in May 2015 that the central section of the railway, between Stratford and Paddington (and including the branch to Abbey Wood) will be controlled. The rest of the line will be run by Network Rail, partly from the Liverpool Street control centre and partly from Didcot, which will look after the sections west of Westbourne Park. This arrangement creates interfaces between Network Rail and Crossrail that require constant communication between the two organizations.

Another source of complexity is the control of the stations. The operator, MTR, will run five of the central London stations while the rest, where there are large numbers of Underground passengers, such as Tottenham Court Road, Farringdon and Liverpool Street will be the responsibility of London Underground though there will be an MTR presence on platforms. On the surface, things are even more complicated: Network Rail will be in charge of the major stations such as Reading (and the surface parts of Liverpool Street and Paddington), while three stations will be the responsibility of the Great Western franchisee and one, Shenfield, of the Greater Anglia franchisee. MTR will be in charge of all the other stations except Heathrow, which is the responsibility of the airport owner.

Despite this split responsibility, there will be strong branding of the Elizabeth Line on the platforms, and clear established standards. Crossrail has, for example, devoted a good deal of attention to the needs of people with disabilities – although at one stage there was no commitment to make all the existing stations on the surface accessible and seven remained without lifts or any step-free access. However, after a long battle by campaigners,

TfL gave in. It committed in 2014 to funding the conversion of four of the remaining stations and, three years later, it announced that it would fund the conversion of the final three – Taplow, Iver and Langley – which are all in the western section of the line. By the time the line opens fully, in 2019 or 2020, all the work should be completed to ensure every platform is accessible.

When a group campaigning for better access to the railways was allowed to examine the new trains, they had only one major complaint. It is one that is likely to attract considerable negative publicity when the line fully opens – namely the lack of toilets. Several suburban lines in London are similarly served by trains without toilets but none offer journey times as long as the 102 minutes between Reading and Shenfield. The campaigners pointed out that Thameslink trains, which, like Crossrail, runs through central London in tunnels, all have full on-board toilet facilities.

Although the trip between Reading and Shenfield will take longer than any single Tube journey, Crossrail argues that journeys from Reading to Shenfield are likely to be the exception rather than the rule, and that the vast majority of people will use the line to travel from outer London to the centre with an average expected journey time of just twenty minutes. Further, since Crossrail uses the slow lines between Reading and Paddington, stopping at every station, many travellers starting out from Reading will take a fast service to London and change. The Crossrail press team says that TfL and the Department for Transport took the decision not to have toilets early in the process. A press officer commented: 'Toilets would displace approximately 600 passengers per hour. At the opening, 34 of the 41 Elizabeth Line stations will have toilet facilities, with a further seven stations having toilet provision in an adjacent building, and all station toilet facilities will include one with full accessibility.' However, the figure of 600 per hour is unlikely to assuage the disappointment of supporters of on-board toilets, as it is a relatively small percentage of potential passenger numbers. There will eventually be twenty-four trains per hour at peak times – making a total theoretical capacity on these trains of 36,000 passengers. Nevertheless, overall toilet provision on the new line will be a considerable improvement on the Underground's past record. The likely explanation behind Crossrail's decision not to install on-board toilets is that having toilets seemed inappropriate for crush-loaded trains in central London and would have added considerable expense, both to the cost of building the trains and to operating them (because of the need to empty the tanks). There would also be reliability issues since train toilets are complex and often go wrong, as regular rail passengers know only too well. There is no chance of the decision been reversed and therefore Crossrail passengers will just have to hold tight in every respect.

12.

The finishing touches

Fitting-out the tunnels was, in many ways, a more intricate and involved task than simply boring them. While the boring was finished in May 2015, having taken just three years, the fitting-out and subsequent testing was a far more time-consuming process, and it was this task which proved to be the undoing of the Crossrail team management. Nevertheless, in 2018, Crossrail appeared to be in control of the process. There were 1,200 people employed in the final delivery stage of the project, but there was still a failure to recognise the sheer scale of the complexity of finishing the scheme.

When in 2018 I met Chris Binns, the chief engineer responsible for the fitting-out, he showed me a list running to several sheets of A4 that set out progress on the various tasks that had to be undertaken to turn the bare tunnels into a functional railway. It was an almost incomprehensibly complex document, divided into heads and subheads, down to the smallest detail. Nonetheless, things seemed to be going relatively well. One advantage in the later stages of the project was that the tunnels themselves could be used to deliver material to the various stations and other worksites underground. In order to be able to do this, the tunnels had to be turned into a temporary diesel-operated railway. Before the rails could be put down, sleepers had to be installed and concrete poured to keep the track in place. The first machines to enter the tunnel were four gantries that were used as track-laying machines. Running with rubber wheels on the concrete sides of the tunnel, they installed sleepers and then sections of track to lie on them. In this process, twenty-eight sleepers were put down with computer-set precision in each phase, and then 108-metre sections of rail were placed on the sleepers and tied to them using clips. Modern railways now use continuous welded rail – which is why trains on modern or upgraded lines no longer make that pleasingly rhythmic 'tagadaga tagadaga' noise which used to send passengers rapidly into la-la land. So, once the rails were in place, the sections were then welded together.

There are, in fact, five different types of track in the tunnels but 80 per cent of it, including all of the track on the eastern side, is 'standard slab track'. Three other kinds of track were used in the main tunnels. Two different types of 'floating track' were installed on 3.5 kilometres of track in areas such as the Barbican and Soho where noise interference from Crossrail needed to be kept to a minimum. Laying track of this kind requires a sophisticated technique that involves jacking up the track and placing it on a series of springs and rubber bearings to minimize noise. A few short sections used 'high attenuation' sleepers, also designed to reduce noise. A fifth type of track was installed in the Connaught Tunnel where, because of the uneven surface on the base of the old tunnel and the restricted height, a type known as 'direct fixed track' was used which, as its name suggests, involved fixing the track directly into the concrete with no sleepers, rather like many tram lines in urban settings.

After the sleepers and track have been laid and positioned correctly by being jacked up, concrete is poured to support them from the concrete train. The train was nearly 500 metres long – the length of four football pitches in the PR language of Crossrail – with different wagons for carrying the materials, providing the power and mixing, pumping and pouring the concrete through a long pipe at the front. In the west, because of the contrasting types of track, a different sort of train, the 'concreting shuttle', which carried ready-mixed concrete, was used.

Among the machines delivered to the various station sites by trains running along the completed tracks were several bespoke contraptions intended to speed up the fitting-out, and designed for a specific function by Crossrail. Several of these machines were bespoke, designed for a specific function by Crossrail. The most sophisticated was the machine that drilled the holes required for the brackets that support the trays that carry the wiring. According to Simon Wright, the programme director who was promoted to chief executive after the departure of Andrew Wolstenholme in March 2018, 'every hole was programmed into the machine's computer so that it knew where to drill every single one. We had the machine specially built for us and ordered it well in advance. So instead of a thousand blokes with a Black and Decker, it's one machine with all the holes programmed in which is far safer as it is not very healthy to drill holes upwards with the cement dust flying into your face, and, of course, the computer is far more accurate.' This machine was used for both main tunnels and was capable of drilling eight holes at a time.

There were, Wright says, 'another four or five machines doing other jobs which were designed specifically by us. It requires a lot of forethought.'

The logistics required for getting material in and out of the tunnels were demanding in the extreme. There are only three portals: at Royal Oak near Paddington, at Plumstead in southeast London (this is actually the entrance to the Thames tunnel but was used as the main way into the central tunnel which starts at Victoria Dock on the north side of the river after the short section on the surface), and Pudding Mill Lane (in east London) and therefore working out how to deliver material in and out of the tunnel required detailed and accurate planning. Plumstead was the main starting point for the trains, though deliveries were possible, too, through the other portals. Once the excavations were completed, the stations themselves – and a few big shafts – offered alternative ways in, but by far the easiest way to transport heavy materials was by using the tunnels.

Most of the work was undertaken during the day, so the pattern was for the deliveries to be loaded onto the wagons at night. The trains then entered the tunnel in the morning, remained in situ along the route during the day, as any movement would have disrupted work in the tunnels, and then left in the evening to be restocked for the following day's work. The wagons had to be loaded with exactly the right equipment for each worksite, as once they were in the tunnel any mistake in the loading could not be rectified. Besides this, any breakdown in the tunnels could wreck a whole day's work for the gangs further up the track.

Indeed, on one of the days I visited Crossrail's Canary Wharf headquarters during a cold spell in January 2018, the management team was frustrated because a fire main had frozen and then burst as the temperature had dipped to four or five below. The tunnels could not be used without a functioning fire main and consequently a whole day's work was lost. Simon Wright emphasized that it was not a matter of planning a day or two in advance, but months: 'Much store is put on thinking tomorrow, next week, next six months. We need to plan for what we need then and some order periods are six months or a year. You can't go to a plant hire yard and get a specialized piece of equipment.'[1] The scale of the operation was quite remarkable. Again, a few numbers tell the story. On average, fifteen trains went into the tunnels and overall, over the course of the near-four-year period during which these diesel trains were used there was a total of 10,000 such train movements. In addition, twenty-one 'mobile elevated working platforms' and eight road–rail vehicles were used every day during installation and testing of the railway systems, and these had to be interspersed with the trains in the correct order. Plumstead was a hive of activity every day.

All these trains were operated by diesel. This was not ideal in the confined space of the tunnels, but battery trains do not have sufficient power and it was not possible at this stage to rig up the electricity required. There was, therefore, no alternative to diesel trains. A whole extra ventilation system, to cope with this temporary situation before the electric overhead wiring could be used, had to be installed to ensure that workers' lives were not put at risk. Diesel trains continued to be used even once the electric wires were fitted because the system was not 'fired up' until May 2018. The trains were only allowed to travel at 15 mph since there was no signalling system to control them.

Once the trains had stopped, the line was put out of use to ensure the safety of the workers.

After the track was laid, all the other systems had to be installed – power cables, electrical connections, signalling and the overhead wiring for electrification. And, of course, the stations had to be fitted-out. The list of what had to be brought in is almost endless: the fans and the pipes connecting them, wire and cable on huge drums, the hugely heavy doors and the rest of the platform barriers, the handrails and aluminium planks for the walkway that is now mandatory for any tunnel this size, the brackets to support the shelving and cables, the signalling equipment, the 'leaky feeders' for the mobile phone coverage, the lighting and the large batteries capable of keeping the lights on for three hours that will probably never be used – plus every single bit of material for the fitting-out of the stations.

Chief engineer Chris Binns is in no doubt that co-ordinating all this was the most difficult part of the whole project:

> The most challenging part of the job is integrating such a large number of systems. It is the enormity of all the systems – building drainage, water services, heat and cooling plant, transformers, switches, crossings and so on – which all need to be functioning and the integration task has been enormous. All these, remember, are different contractors. There are all the station systems, too, such as barriers, communications, PA systems, CCTV with every camera available to the British Transport Police. And details, like the station contractors have to ensure there is room available for those fitting these systems. Everyone who goes on site has to be made familiar with the safety rules and our procedures. In 2017,

we brought in all the contractors at one stage to help them work together.[2]

There was a set order for the installation of the equipment: 'So the wire went in, then cable trays for low voltage and high voltage feeder systems [the low voltage is for all routine electrical purposes such as lighting and powering equipment, while the equipment for the high voltage, 25,000 volts, is for the wires – the catenary – that power the trains], then the handrail and fire mains – all the longitudinal systems.' The signalling equipment, too, had to be brought in as each station has a signalling equipment room which controls the system in the local area.

For Bill Tucker, the American who was in charge of the construction of the central section, the hardest aspect was dealing with contractors. He reckons that the way the industry has changed over time has made the job of project management far harder. He provided a fascinating insight, which has great resonance for other projects, into the way that the construction industry has changed during the eighteen years he has been working in the UK.

Tucker explained that the Thatcherite ethos of contracting out work as much as possible along with the Fordian notion of breaking down tasks has transformed the industry, but not necessarily for the better. Outsourcing tasks, even major ones, has become the norm for companies, greatly reducing the number of people in their labour forces for whom they are directly responsible. These changes have combined to produce a situation in the construction industry that is radically different from 2000, when Tucker first arrived in the UK:

It used to be that firms called themselves construction companies and they directly employed lots of people, as well as a few contractors. Now they call themselves contractors, and it is the contract, rather than the construction, which has become the focus of the job. The day they sign the contract, they come with fifty quantity surveyors and lawyers, who will go over the contract and say, "we need this and that variation because it is not specified and so on". That is not the way it used to work when a construction firm would simply do the job and settle up at the end of it by sorting out the variations.[3]

Tucker explained, too, how the pattern of employment has changed. Whereas in the past the construction companies employed large numbers of people who had a long-term commitment to the firm, now they contract out most of the jobs to labour-only subcontractors: 'Consequently, they do not have the long-term loyalty which is the norm and it is therefore much more difficult to ensure that people provide the right quality of work.' That was one of the reasons why Crossrail felt it was very important to establish that everyone worked for the project, not for their own individual company. There is one exception: Laing O'Rourke, which took out several contracts for Crossrail work in the traditional way, retaining a large directly employed skilled workforce which moves from job to job. Interestingly, Laing O'Rourke proved ultimately to be by far the best of the various Tier 1 (leading) contractors.

Tucker went on to describe how the final fitting-out was handled differently from other Crossrail contracts:

There are three different approaches. For the tunnelling, we actually commissioned contractors to come up with detailed designs, to show us precisely how it would be done and leaving a lot of the precise details up to them. For the stations, we did much more specification, actually saying how we wanted the station to look, but for the fit-out, it was different. We pretty much said, "this is what we need" – and then left it to them how to do it.

Much of the work on each different system involved in the fit-out could be carried out independently, but Tucker provided an excellent example of how different aspects can impact upon each other:

We went through an elaborate process of identifying all the inputs and trying to work out the interaction between them. Take, for example, a fan. Being able to identify the size of the fan required is complicated. What's the size of the tunnel? What are the aerodynamics going through the tunnel? What is the configuration between the tunnels? All those considerations and several others have to be taken into account to know what type of fan to specify. And this in turn determines the size and weight of the fan, which will also then have an impact on the design of the station.

So far, so good. But then a feature of the trains that had hitherto hardly seemed significant caused an unexpected ripple elsewhere in the project that had widespread knock-on effects. The trains Bombardier designed for Crossrail were lighter than had originally been envisaged, and occupied a smaller amount of space (the technical term is 'envelope') when going through the tunnels. Tube trains, as their passengers know well from the familiar rush of wind that precedes their arrival, do much of the ventilation of the tunnels themselves, by pushing air ahead of them. It was expected that the Crossrail trains would do the same, but a slightly smaller and lighter train pushes less air, because it fills up a smaller proportion of the tunnel envelope. Tucker elaborated:

That changed the whole calculation. Bombardier's solution was more economical in terms of maintenance, but it meant we needed bigger fans. The fans had to do more because the trains did less, and these bigger fans caused more vibration and noise, and therefore needed more noise abatement and better insulation from the vibration. So one little change can have a ripple effect on station design, acoustic mitigation and the design of the fan itself. And that is just one example that when you do the development design, the assumptions may not be correct. The parameters put in by the engineers can change.

Tucker had more to say about unforeseen side effects:

Another example is communications. All the communications run from the control centre at Romford through a system called SCADA [Supervisory Control and Data Acquisition – a fancy term for a system that uses computers and radio communications to provide the overall control of the system]. So assumptions are made on the volume of data and the required number of points of interface that are going to be needed to link all the pieces of equipment in the stations and on the track.

But if, he continued, 'a designer picks a different pump or switch gear, it might then require a different number of input or output points. So if at Paddington a change like that is made, it might have implications for what is needed twenty miles away at Romford.'

Variations of this type, which inevitably have a financial impact, demonstrate that not all increases in cost are caused by mistakes or greedy contractors; some derive from unexpected effects of the myriad interfaces between the different systems that make up a railway. Indeed, Tucker stresses: 'The majority of change that we have dealt with at Crossrail has primarily been of interface changes in the design which then require physical change.'

In addition to the tragic death of René Tkácik, which hung heavily over all those involved in the project, there were, inevitably, several other accidents, albeit with less tragic consequences but more significant in terms of the project's progress. The most significant of these was in November 2017. When the main power electricity was turned on for the eastern section of the tunnels, there was an explosion in a transformer which linked the power from the grid to Network Rail at Pudding Mill Lane in east London. When the transformer, a complicated piece of equipment, was connected, the insulation started melting and after several days it disappeared entirely, causing a short circuit involving a 25,000-volt cable. Binns believes there was a design flaw:

It was a small transformer but a big bang. It was very spectacular unfortunately, but luckily no one was hurt. It was just a measuring transformer, very small but it had around 8,000 amps [a hair dryer typically uses 10 amps, a kettle 13 amps] for 0.18 seconds, so it was very short amount of time but a huge amount of energy. It literally blew up. We made sure that would never happen again, so they are fused in a different way now and testing procedures have been changed. We learnt some hard lessons from that.

This delayed the 'energization' of the electric supply for three months and crucially delayed testing. As we will see in the next chapter, it was a key moment for the project as it was a warning that the deadline was impossible to meet. But it also proved to be a missed opportunity.

By the summer of 2018, when the first edition of this book went to press, the testing regime, which seems almost endless, was in full swing. One train into the tunnels at slow speed, then at line speed, then with the doors being operated, then more trains and so on, to test every system, 'making the service fit for service', as Terry Morgan put it. Of course, it was not just the trains. Every aspect of the stations has to be tested, such as making sure that when the fire alarms go off the barriers swing open and stay like that. The ventilation system is an organism in itself and has to react in the right way to smoke. All the platform doors have to work, and work every time. Chris Binns says that

awareness of making maintenance easy has been a constant and difficult strand of work. There will only be four hours to keep the railway in a good state every night, and we already know how many closures we will need each year and for how long. And if a decision is made to use Crossrail at night, as has happened with parts of the Underground and, more recently, the Overground, it should be possible to support it.

By mid-2018, 200 people had already been recruited as the permanent maintenance team.

However, despite all the attention being paid to the fit-out and the integration of the various systems, the schedule was unrealistic even then. The fit-out involved a myriad different systems connected with state-of-art (always a dangerous expression for new projects) software with all the work being carried out on building sites in the centre of London that were difficult to access. In truth, in 2018, the project was nowhere near finished.

13.

And another one?

A part from a few updates and corrections, the preceding chapters tell the story of the situation in the summer of 2018. I had spent much of the early months of the year in the Crossrail offices, which were then in Canary Wharf, discussing the project with senior executives, and going on site visits to several of the stations, and even for a walk along one of the tunnels. Even though I was a bit sceptical of the scale of the work needed to complete the project in time for the opening date of 9 December, the task seemed just about feasible. The team would muddle through, overcoming the various obstacles and challenges. They had done so before. The phrase 'on time, on budget' was such a strong mantra, and one repeated by all the top managers with greater or lesser degrees of conviction, that it seemed unlikely that they were blagging it.

I had, to some extent, drunk the Kool-Aid. These were highly competent managers who seemed to be keeping on top of the task – just about. Moreover, there was an element of Stockholm Syndrome. I had spent a lot of time with them, and they were a likeable group of people. I had been impressed by the collegiate atmosphere which seemed genuinely to permeate through to the teams working at the coalface. Although it seemed to me that those teams were rather too large given the proximity of the schedule opening date. And that was my first big hint that the optimism pervading the whole project was misplaced. In particular, on a visit with Terry Morgan to Paddington station in June 2018, I was struck by the number of people working there and the number of tasks that remained to be completed. There were at least 500 people on site and nothing was finished. There were orange-jacketed men (and the occasional woman) everywhere, beavering away at a multitude of different tasks ranging from fitting the escalators to installing various pieces of electrical equipment. The amount of equipment lying around covered in polythene and waiting to be fitted filled several of the various rooms that are part of every station. The control centre – clearly nowhere near completion – was a mess of wires and unwrapped computer terminals. A few platform doors had arrived, propped against a newly concreted wall, but clearly there was much work on the platforms to be done before the fit-out could be started. This was very much still a building site rather than a station that would have thousands of passengers passing through it in just six months time.

And this was not the station that was most behind schedule. Bond Street, as Morgan admitted at the time, was deeply problematic with scope changes such as the need for an additional ticket hall (a quaint name given that tickets are now a few lines of code on a piece of plastic) and there were difficulties with the contractor, Costain Skanska. The Crossrail PR team were understandably reluctant to let me see it. Whitechapel, too, was off limits. This was a particularly complex project which necessitated working around the existing Overground and Underground lines, as well as a large sewer; it was, according to all reports, still a mess. There was talk of the train service starting up without stopping at these stations, but there were still safety considerations that needed to be addressed, notably that the stations would have to be sufficiently completed in order to provide an emergency exit should there be an incident on the railway.

Following the visit, I asked a senior manager, promising it was off the record, about the chances of the project opening on time. He hesitated, and then said '80 per cent' stressing that they were intent on not doing a partial opening with an unreliable railway. It had to be good enough for the Queen to perform the ceremony and for the public to travel with the expectation of a safe and reliable journey. In July, with the expectation that the line would still open on time, Crossrail had secured extra funding of £590m provided jointly by Transport for London and the Department for Transport. That seemed like a ransom fee paid in return for a 'get this done at any cost' commitment from Crossrail.

But it was all a pipedream. By then, Terry Morgan was beginning to realise the game was up. Along with other board members, Morgan went to see Sadiq Khan, the mayor of London, on 29 August to give him the bad news. The July 2018 bail-out would be the first of several and the Queen would not be dropping in on 9 December. Moreover, it was impossible to provide a firm date for the opening of the line as Morgan told Sadiq he needed a couple of weeks to work out a new timescale for the opening. Morgan faced another difficult meeting the following day when he went to advise Chris Grayling, the Transport Secretary, of the delay. Ever the politician, Grayling unhelpfully responded that it was not the concern of central government. The formal announcement of the delay was made by Crossrail on 31 August.

In fact, it seems clear that there was unnecessary procrastination about the announcement. On 26 July, Morgan, along with his chief executive, Simon Wright, gave a presentation to senior managers and politicians at Transport for London including the mayor, the deputy mayor for transport Heidi Alexander, Mark Wild, who was then in charge of London Underground, and Mike Brown, the Transport Commissioner.

The message from the Crossrail team's PowerPoint presentation was very clear – and pretty bleak. They put forward three possible scenarios for when dynamic train testing could start: there was a 10 per cent chance that it would begin in November, pushing trial running and actual Elizabeth Line operations back to somewhere between February and May 2019. More likely, if testing took place at the same time as other construction work, there was a 50 per cent chance that the Elizabeth Line could open between April and June 2019. However, if work on the routeway continued to be delayed the line would open between May and August 2019 – but all these scenarios depended on the testing not throwing up major issues. The message seemed to be clear. The December 2018 deadline was not going to be met.

According to the highly-regarded and well-informed online *London Reconnections* journal, none of this should have been a surprise to Brown and his team as Siemens had written to him directly in early 2018. In that letter they had warned that, from their perspective, the train programme (for which they were heavily involved in signals integration) was seriously behind schedule... Siemens had warned Mike Brown that even at that stage, as things stood, it was unlikely that they would be able to commission the signalling and pass the trains through testing in time for a December opening.

Yet somehow this all fell on deaf ears. Admittedly, for the mayor this may have been news, but it seems remarkable, since Crossrail had all but admitted the December timetable would not be met, that the Mayor and TfL did not make the delay public until a month later. As a result of the differing interpretations of what happened at this and subsequent meetings, the story in the media quickly degenerated into a row over the precise timing of when and what Morgan had told Sadiq about the delay, which really was a side issue given the enormity of the impact on London and, specifically, on the finances of Transport for London. Morgan is adamant that he had informed TfL of the situation but insists that his emails to Brown had been watered down by the time they reached the Mayor. *Crossrail Delay*, a report published in April 2019 by the London Assembly, confirms this. It was expected that the line would open some time in 2019, but no specific date was given and the Crossrail team went back to the drawing board, reluctant to commit itself.

The principal mystery was, of course, how could the team have believed that it would deliver the line by 9 December 2018 when even a cursory investigation should have revealed that there were insurmountable difficulties? The issue is made more complex because it was not just the Crossrail leadership team that had been fooled or, indeed, had fooled itself. A long line of people and organisations seem to have been taken in by Crossrail's bullishness. For example, in 2016, Chris Croft, a partner in KPMG's Infrastructure Advisory Group, wrote about the project in the organization's in-house magazine, *Foresight*, boldly stating in the introduction that 'Integrated sponsors, an independent delivery body and strong governance have made this high-risk undertaking a potential text-book case in mega-project management'.[1] His article went on: 'Projects of this magnitude are often plagued by inefficiencies, delays and overspend. Yet, Crossrail has largely managed to avoid these excesses.' KPMG were not the only ones to be overtly positive at a time when the project was already beginning to fail. The Infrastructure and Projects Authority, part of the Cabinet Office, gave Crossrail an 'amber green' rating in 2017 in an unpublished report, pretty much endorsing the programme as it was set out. Various internal reports by both the government and Transport for London had failed to sound any warning bells about the project's deep-rooted difficulties. As late as December 2017, TfL announced it was seeking 'launch partners' for the opening of the Elizabeth Line, offering 'a unique opportunity that will align with this historic moment for London'.

Doubts were already being raised from one crucial source, but these warnings were not being made public. Jacobs, the consultancy which had the role of the government's Public Representative on the scheme, essentially playing the role of watchdog, raised questions with the Crossrail team about the lack of progress and expressed the possibility that the deadline would not be achieved. In its heavily redacted report, which was produced in January 2018 but not made publicly available until December 2018, Jacobs stated:

There is a risk that the train software will not be sufficiently advanced to support EDT [Early Dynamic Testing of the trains in the tunnels] by the end of February 2018. As a result there is a significant risk to formal Handover. CRL [Crossrail] continues to prepare a revised MOHS [Master Overall Handover Schedule], expected in February 2018, to demonstrate how Crossrail completion can be achieved.[2]

Five months later, in May, the warnings were starker:

The Master Operational Handover Schedule (MOHS) remains highly ambitious. There is little or no float available ahead of Zones 3 & 4 dynamic testing, or to allow sufficient completion of works ahead of Trial Running.

(Dynamic testing is the first stage when trains are allowed to be tested without any passengers in the tunnels while trial running is when they operate to a fixed timetable with the possibility of carrying passengers to test safety and other issues.) A new system of blockades had helped, according to Jacobs, but 'this improvement is not so evident in the installation, testing and commissioning works associated with Stations, Portals and Shafts, where significant completion and integration challenges remain'.[3]

Clearly the writing was on the wall and by the summer the handover schedule had fallen apart. Simon Wright, who became chief executive in April 2018, was forced to create a new 'MOHS', but as Mark Wild recalls, 'it became apparent that we were not going to make it'.[4] Nevertheless, it was still another three months before the leadership team threw in the towel.

The MPs on the Public Accounts Committee enquiry into what went wrong were clearly dissatisfied with the Department's response to their questioning about the failure to heed the warnings. Quoting Jacobs' May 2018 report, the MPs pointed out that 'the operational readiness assessment within the report notes that 16 readiness tasks are rated as 'red' and a further 5 are rated as 'amber'. None are rated at 'green'. The report summarised that the schedule 'remains highly ambitious' and there 'remains a high risk that [the December 2018 opening date] will not be achieved.'[5] The MPs could not understand why the Department, which was receiving these monthly reports from Jacobs, had not pressed Crossrail more strongly on its ability to meet the deadline:

The Department was unable to explain how it reconciled the Project Representative's bleak assessment with other information it was getting from Crossrail Limited that the programme would be delivered on time. The Department argued, unconvincingly, that the Project Representative was a small team undertaking deep dives into small aspects of the programme, compared to thousands of members of staff in Crossrail Limited. It told us that while it was aware of delivery risks to the December 2018 delivery date, it expected these to be a matter of weeks not months and the scale of the problems emerged far faster than it expected. It recognised that there had been warning bells in March and April 2018 because of projected cost increases, but given the scale of the programme, it had believed these were not as major as the problems that crystallised over summer 2018.[6]

Neither the Department nor the MPs thought that there was a deliberate attempt to deceive the government, but, rather, that there was a collective mindset which did not recognize the possibility of failure. The Crossrail executives I met during the first half of 2018 were, in the words used by politicians who are about to lose election campaigns, 'cautiously optimistic' despite the statements they were receiving privately from Jacobs. If they did consider the possibility of a delay, it would be in the order of months, not three years.

One possible explanation behind the reluctance to admit the delay was inevitable is offered by the anonymous writer in the *London Reconnections* journal:

> As soon as you concede there is a delay contractors might regard themselves as 'off the hook' for any delay caused by them and not proceed with such haste. A pre-emptive announcement of a delay could make litigation against contractors more difficult to win in court. And the claimant in court has to show that he did everything he could to mitigate his losses by trying to make up time. Keeping quiet to senior employees about the full extent of the problems and the delay caused could be for a similar reason to the one involving contractors. It could lead to Crossrail employees who believe that they are not on the critical path perceiving that the sense of urgency has gone away.[7]

While that may not be a complete explanation for what, on the face of it, seems almost inexplicable, it makes sense when combined with the various other possible elements of the story.

So strong was the confidence of the Crossrail team that

during the first half of 2018 senior people were being 'let go' on the assumption that the task was nearly complete. The chief executive, Andrew Wolstenholme, left in April and so had several senior project delivery people along with vast numbers of staff lower down the hierarchy. Overall, during the first half of 2018, around 1,000 people who were responsible for the delivery of the project had been demobilized. The tone of the press release issued during this period is revealing: 'Crossrail Limited today announced that as part of the planned demobilisation of its delivery and leadership team, Chief Executive Andrew Wolstenholme OBE is stepping down after seven years at the helm to take up a new role in the private sector.' It went on to quote Terry Morgan: 'Construction of the Elizabeth Line has entered its final stages and during the coming year we will be handing over the completed assets to Transport for London, who will lead the final testing and commissioning phase ahead of the railway's opening in December.'[8] As further evidence of overconfidence, Wolstenholme was not really replaced as the role of chief executive was given to Simon Wright, the performance director, who also retained his current job. Neither of the sponsors, Transport for London and the Department for Transport, seemed concerned with progress. Indeed, so confident was Grayling in the soundness of the project that on 1 August he appointed Morgan to be chair of HS2, an even larger megaproject which was already under fire because of cost overruns and delays in construction.

There are two issues here. The fact that a project of this magnitude ended up being delayed was hardly remarkable given its sheer scale and pioneering nature as Britain's first fully digital railway. The truth was that a timetable set rather arbitrarily in 2011 for a project of this magnitude had never really been feasible. More remarkable is the fact that it took until just four months before the opening for the leadership team to recognize the target would not be met. And what makes this even more mystifying is that the deadline was not missed by a matter of weeks, or even months, *but by more than three years*. Tony Meggs, who replaced Morgan as chairman in December 2018, suggests that the only possible explanation lies in the culture of the organization and the familiar phenomenon of the emperor's new clothes: 'Crossrail was the best project in Europe, indeed in the world, until it wasn't. However, if you got right down into the trenches, people pretty much knew that it was not going to happen but that info got filtered as it found its way up and the senior management were only presented with what they wanted to hear.'[9] Indeed, Mark Wild, who took over as chief executive in November 2018 reckons '99 out of 100 people at Crossrail knew that the timetable could not be met but did not want to be the first to put their hands up'.[10]

Meggs suggests it was a mixture of overconfidence and an obsession with meeting the deadline:

Somehow, the climate in the organisation was such that people were unable to tell the truth because the leadership was so fixated on December 9th that they literally could see nothing else. They had overcome lots of problems so far, they had bags of self belief, and they did not want to let the air out of the system. And indeed they wanted to keep the pressure on.[11]

However, at the end of the day, there was a fundamental

mistake, repeated by everyone involved in the project as Wild puts it: 'They really did not have any idea of what they had left to do. Assurance, validation, verification and handover just got squished and squished until there was no time left to do it.' For him, the decision to allow so many people to leave early was the most difficult to understand: 'When you look back, of all the things that were inexplicable, it is that which leads you to think the leadership and the board were so confident in their ability to complete.'[12]

According to Meggs, Crossrail's management were afflicted by a collective deafness to reality:

the stakeholders... at the senior level were all convinced it would open on time. Contractors would present more realistic schedules but they were rejected by Crossrail saying "that is an unacceptable schedule". There was huge tension in the system, with people who wanted to deliver the railway on time but couldn't hear – didn't want to hear – that it could not be delivered on time.[13]

During my visits to Canary Wharf, I was told repeatedly that the scheme was 93 per cent complete but Meggs suggests this was based on the wrong metrics: 'It meant 93 per cent of the money had been spent, not 93 per cent of the work completed.' The reliance on the wrong metrics by the 2018 leadership team was a key refrain from those assessing what went wrong. Wild, who was on the Crossrail board as the Transport for London representative from September 2016 – and is conscious that he should have been able to see the wider picture – also mentions that they were basing their decisions on the wrong factors:

The truth of the programme was not in the metrics. That's the key point. For the whole of 2017, the key risk that was accumulating really fast was not in any of the metrics. The level of 'completeness' was over-represented because we were using a measure that was mathematically accurate, which was the wrong thing. They should have measured what was in the document control systems, the assurance system and signed off. Things that were really done. They weren't doing that. They were measuring things that the contractors had claimed.[14]

He acknowledges that Jacobs did express doubts but he feels they were not sufficiently strong in their concerns: 'Their reports of late 2017 and early 2018 said things like "there are signs of progress stress" but not "you are going to overrun by three years and £3bn". So I think the sponsors [TfL and the Department] did the best with the information they had. I honestly think it is people like the board and me who should look at themselves.' He added: 'I have a problem as I know these people and they are good professionals. This was a great team and you can't personalize it to one person as it is systemic. There is no doubt that victory was declared a bit early at times.'[15] Certainly if the Jacobs reports had been made available publicly as soon as they were completed, more questions – and better informed ones – would have been asked both by the civil servants overseeing the programme at the Department for Transport and Transport for London and, crucially, by the respective politicians.

Mark Wild's view is that 'the goose was cooked' as early as September 2016. 'The opening date wasn't achievable as in 2017 pressure increased, with mitigations, interventions and then the Pudding Mill Lane explosion.' All the key players now agree that this was the point where a delay could have been announced with little comeback. A faulty transformer blew up at Pudding Mill Lane on 11 November 2017, causing the testing programme to be put on hold. At the time Crossrail played down the seriousness of the incident rather than acknowledging it, which resulted in a series of consequential delays. The writer in the *London Reconnections* journal put it well in an article published in January 2019:

Crossrail – and TfL by abstraction – missed a golden opportunity to play a "get out of jail free" card. Publicise the delays, emphasise how completely unexpected the problem was and how it was (genuinely) down to one of their suppliers, and suggest that because of this the opening date was now in doubt. This didn't happen.[16]

Terry Morgan admits that it was a crucial missed opportunity. Recalling the incident in an interview in August 2021, he told me:

If I look back, the time we should have waved a white flag was when we had the [Pudding Mill] explosion. Instead, it turned out to be a heroic exercise to try to mitigate it. It's what we did all the time. It's going to be late, mitigate. We kept on trying to fight to keep it going. It was a simple incident but as a consequence we could not run the railway.[17]

That was a part of a malign pattern – the refusal to acknowledge that the deadline could not be met, which in turn led to attempts to patch up holes that simply could not be filled.

A typical example of this problem was that stations were being built before the design process had been completed. Design changes were being made on site which then had to be reported back so that they could be included in the final design overview. Chris Binns, the chief engineer, reckons that there were some 2,000 design actions during construction of the stations – in other words actions based on decisions made on site rather than in advance – and these all had to be documented. While many of these changes were minor, some involved considerable work. Binns says that the method used by workers to climb off the tracks had to be completely changed throughout the network: 'We started to get feedback from RfLI [Rail for London Infrastructure, the Transport for London subsidiary in charge of the track] that our idea of installing stirrups to give people a leg up out of the walkway was unsatisfactory, which was correct in the modern world. We had to replace them with ladders – eighty of them – all the way along the track.'[18] In November 2018, Binns recalls that there were still 500 design reviews for which documentation had to be produced, some of which were safety-critical. All the more baffling, then, that the impossibility of meeting the deadline was not recognized earlier.

The National Audit Office report into Crossrail published in May 2019 is comparatively restrained in its criticism but states clearly that there was a failure by management to accept the inevitable. The report suggests the causes of delay went as far back as 2015 and should have been apparent well before 2018. Meggs backs this up: 'If you look at the records, from 2016/17 there was slippage. They said the tunnels were on time and on schedule but that was not really the case. If you overlay the original plans with what happened, there was a continuous process of rescheduling without ever moving the 9 December opening date.'[19]

The NAO report recognises the problem caused by the fixation on the 9 December opening date, which was set as far back as 2011. While the NAO was clearly right to highlight this issue, it is remarkable that all the other investigations into the project by the likes of KPMG, the Infrastructure and Projects Authority and innumerable other investigations either missed this crucial point or ignored it. The continued emphasis on the 9 December date was not only misplaced but extremely costly and, ironically, resulted in a later opening date than would otherwise have been the case. According to Meggs, interviewed in the summer of 2021, 'had we recognized that more time was needed in 2016 or 2017, we would have been able to stretch out the schedule and the project might be finished now – we lost a year because of the failure to do that'.[20] A further failure was the lack of any contingency planning for what should be done if the 9 December deadline were not met. Failure was simply not considered; the consequent absence of a Plan B undoubtedly led to a lengthening of the delay.

The NAO reported that 'The delivery approach, delays to some contracts and the decision to set and then stick to the December 2018 opening date, led to increased compression in the programme and increased risks'. The next sentence is telling: 'A number of stakeholders we spoke to expressed the view that the Crossrail Ltd executive team recognized the challenges but believed this was an exceptional team capable of delivering exceptional results and overcoming these challenges.'[21] The NAO's words highlight the bullish, 'we can overcome' attitude of the management, which I experienced personally during my 2018 visits to Crossrail. This blinkered way of thinking closed management minds to the consideration of alternative options. Moreover, at root, the 'can do' approach masked the fact that Crossrail was a more complex enterprise than any that had been undertaken before. Ultimately, Crossrail's leadership were guilty of a failure fully to understand their own project.

Perhaps the most intractable of Crossrail's problems was the rather nebulous but absolutely fundamental concept of 'systems integration'. While all railways need to be integrated in that the trains have to follow signals, run on the tracks, stop at stations and operate to a timetable, these various tasks were, in the past, provided discretely and could function independently of each other. On Crossrail, all the information is digitized and fed through a control and monitoring system known as SCADA (Supervisory Control and Data Acquisition software). Moreover, software is at the heart of every system, whether it is the tunnel ventilation system or the light switches in a cupboard at Whitechapel station. All this information is monitored through the control system which is operated from the second floor of a Network Rail building by the side of the tracks in Romford.

The integration of these systems, involving the dovetailing of some 60,000 elements, is probably the most complex task ever attempted on Britain's railway network. Moreover, the interaction between the various elements of the system needed to be understood right from the start of the construction process. This was, however, not always the case, and was a cause of delays. For example, the stations were each built through separate contracts, but the communications between them and with the control centre had been set up under a contract known as C660, which was won by Siemens. The work entailed designing and installing the communications and control systems within Crossrail's central section, including CCTV, public address systems at stations, customer information displays and emergency services radio systems as well as communication with the Romford control centre. The existence of as many as thirty-seven major contracts – as well as hundreds of sub-contracts – which were often interdependent in all kinds of unexpected ways, made for an exceptionally complex situation, with massive potential for problems of integration. Mark Wild gives the example of station fit-outs: 'No one measured the interconnectedness between, say, contractor A and contractor B. So stations could not be completed until the CCTV system was put in and that was done by the Siemens C660 contract, not the station contractor.' He explains that there was simply no way of measuring these interfaces: 'Not only were they using the wrong measure, but even if they were using the right measure the effort of integration was not quantified in such a fragmented environment.'[22]

Binns also highlighted low levels of productivity, a problem greatly exacerbated by the interaction between the different parts of the wider system. Productivity was measured as a percentage of the work completed compared with the programme set out for that day. At best it was around a third; in other words two-thirds of the tasks planned for each operation were not completed. Lack of integration between programmes meant that one contractor had to wait for another to finish a particular job before being able to carry out their work. Bringing all the components of the railway into the SCADA system proved to be a much greater task than had been envisaged. Only fourteen weeks had been allocated to integrate the electronics on the ventilation shafts within SCADA, but in reality almost twice as long was needed. As for the stations themselves, integration would take a whole year as myriad devices – light switches, cameras, alarms and others – had to be connected to the system. Managers assumed that the stations were virtually finished because their components were physically in place, but there was still considerable work to do before they could obtain the safety assurance needed to allow passengers in.

The installation of cameras and alarms provides a sobering example of the scale of the task. It was discovered that there were places that could not be seen on camera and door contacts which had not been fitted. Seventy-six more cameras were needed, bringing the total up to a staggering 1,599 (a figure that does not include the many cameras on existing London Underground stations). On top of 1,000 door alarm contacts an additional eighty-one needed to be fitted.

The most important interface is, of course, between the train and the signalling. Unsurprisingly, this proved to be the most troublesome element of the integration process, not least because – as we have already seen – Crossrail makes use of three different signalling systems. The trains have to be adapted for each of the different systems, but also to cope with the interfaces between them as they switch from one to another. The Pudding Mill explosion had led to a delay to the start of train testing, which only began at the beginning of 2018. From the outset Crossrail encountered significant problems getting the signalling equipment on the trains to communicate with the trackside-based equipment and with other trains, both of which were essential for safe running as in the tunnels the trains are entirely computer-driven. In the Heathrow tunnel, where the ETCS (European Train Control System) is used rather than the CBTC (Communications-Based Train Control) that is deployed in the main tunnels, it proved especially difficult to coax the 345-class trains into communicating with the signalling system. Modern signalling systems based on radio signals are susceptible to interference from other electronic and electric devices – a problem that is very hard to mitigate. While these systems are bench tested – in other words tried out offline in laboratory conditions – it is only when the trains run through the tunnels that these glitches and interference patterns emerge.

While the Pudding Mill Lane incident held up the start of the process of bringing the trains into both the Heathrow and main tunnels, there were already a host of errors in the signalling system which needed attention. Binns recalls that in October 2017, just before the incident, 505 cfxs (errors) in the system had been dealt with, another 206 being worked on and a further 60 whose cause was still unclear. While many of these might be minor – such as a faulty display on indicator – some are safety-critical and all have to be addressed. A huge software drop which took place as the revised edition of this book was being finalized in October 2021 included the correction of around 200 errors, some of them relating to quite major malfunctions such as trains stopping in the wrong place. This was something I myself noticed on a visit to Whitechapel station in July 2021 when the station was officially being handed over from the contractors to Transport for London. The test trains running through every five minutes were all stopping a foot or two away from the point where the middle of the train doors should align with those on the platform. Binns assured me later that this would be remedied in the big software drop, but it seemed like a pretty fundamental error so far down the line and three years after trains had started being run in the tunnel. A second major software drop was scheduled for Christmas 2021 which would determine whether the hoped-for opening date of February 2022 would be met.

The scale of the work undertaken after December 2018, outlined in the next chapter, is quite remarkable given that the deadline was not abandoned until just four months prior to that date. In this chapter I have set out a lot of different reasons why delay to the original opening date for Crossrail was inevitable, but the explanation for the ostrich-like attitude of the Crossrail leadership in 2018 remains as mysterious as the fate of the Marie Celeste. The physical aspects, such as the scale of the task and the failure to recognize the difficulties of integrating the various systems, can offer only a partial explanation for this astonishing lacuna. The rest is down to the psychology of those leading the project and the conditions under which they were working. The reluctance of staff lower down the hierarchy to alert the leadership to the scale of the problems, the phenomenon of the emperor's new clothes, the fear of failure and retribution, the ambivalent position of contractors who stand to gain from extra work, the lack of clear oversight from the project sponsors, the tin ears of the politicians and other stakeholders and the fear of failure all played their parts in the corporate mystery-drama that was the Great Crossrail Delay of August 2018. However, while these factors contributed to the situation, a large part of it can only be explained by the ubiquity of human failings.

Christmas 2018 was the nadir of the project. Wild had just

got his feet under the desk, but as yet he had no schedule for when the project would be finished. To add to the sense of uncertainty, over the holiday period Crossrail moved offices from Canary Wharf to Olympic Park four miles to the north. Wild describes this as a 'comedy moment when we were in crisis but people were still trying to find where the pencils were'. But there was little that was comic about the situation that Wild had inherited:

> it was chaotic since the heads of the organization had left and we did not have any structure really: the delivery organization was new, the project control was new, and the project had become chaotic in terms of data and knowledge. There was not enough information about what had been done and there was a lack of data because lots of the project control organization had been let go. So the biggest problem I faced was that I did not know what was happening.[23]

2019 was not going to be an easy year.

14.

Rescue

When he arrived as chief executive in November 2018, three months after Terry Morgan's fateful meeting with Sadiq Khan, Mark Wild thought there was just six months' work left to do. He was soon disabused. Even then the sheer scale of the task of finishing the railway was nowhere near understood. It makes the errors of the board and the senior managers in underestimating the task in the first place easier to understand. This was uncharted territory: Britain's first digital railway; a very challenging working environment in the centre of Europe's biggest city; a complex management structure with two organizations overseeing the project; and, above all, a 'system integration' process like nothing that had ever been attempted before. They knew they had a mountain to climb; the only problem was, they thought it was Snowdon when in fact it was Everest. As Rhys Williams, Crossrail's head of mechanical, electrical and public health put it: 'we've built two railways – one real, one virtual reality'[1] and it was the second of these that no one quite understood.

At a briefing on 11 December 2018, Mark Wild, just a month into the job, attempted to set out the state of progress of the project to the London Assembly. It was already clear that hopes of a relatively short delay before opening were fast evaporating. Wild, who was moved across from his previous job running the Underground was widely recognized to be the right man – indeed possibly one of the *only* men suitably qualified – for the job. John Bull, the editor of *London Reconnections* argued persuasively: 'It is hard to think of a better, more-suited, more-qualified person to take over.'[2] Quietly-spoken but extremely sharp and very experienced in dealing with big projects, Wild is not someone who is afraid of making radical decisions and, crucially, he vowed to ensure that everything Crossrail did was more transparent and open to scrutiny. Had it been so previously, then the deadline debacle would never have happened. Openness was crucial to regain the confidence of both the public, which had lost faith in the project, and the politicians, whose support it still needed. Wild remained, however, very reluctant to commit to a new opening date, despite the pressure on him to do so.

To the Assembly, Wild provided a lengthy explanation of the various problems with the signalling systems, both in transitioning between systems and the difficulties in adapting them to the specific complex needs of Crossrail with its connections to the national rail network and to the Heathrow tunnels. Indeed, testing of the trains, at this stage, seemed to have hardly progressed. The trains had been delivered late by Bombardier (Morgan and Wild claim it was three years late but that depends on how one defines a finished train as so much of it is software) and the Pudding Mill Lane explosion had further held up testing which had started – belatedly – in May

2017. Wild explains that they had expected they would need something like five iterations of the software to get things right and that this would happen in the middle of 2018, but in fact by then they were on the equivalent of version 21 of the software development and it was still nothing like the finished product.

Wild admitted later, 'we had this delusional thing, as the assurance tail would have killed us anyway'. In other words, the lengthy assurance process would have pushed Crossrail way beyond December 2018 – even had the project not been beset by other problems, the most significant of these being that much of the testing, which was not being carried out under proper conditions because of the time imperative to get it done simultaneously with other work on site, was invalid. Another major issue was that the Train Control Management System – in other words the controls on the train – was having difficulty communicating with the ETCS in the Heathrow tunnels and at this stage the 345s were not actually able to enter them. The problem was highlighted by what Wild later told me was a fundamental mistake in the procurement process. The trains had been ordered by Transport for London and were therefore its responsibility rather than Crossrail's, thereby creating yet another interface. So it was that Howard Smith, whose job was to commission the trains, reported to the TfL hierarchy* rather than to the Crossrail board.

Wild emphasised that the idea Crossrail was 93 per cent finished, as his predecessor had claimed, was based on a lack of understanding of the metrics.

---

* By a curious coincidence, after 2016 he reported to Wild, who became MD of London Underground in June of that year.

There was no question that there was merely 7 or 8 per cent of the project to complete. In the supply chain, they were simply counting what they had done but that did not even include the approval process. To get a green field safety case completed, there were 230,000 safety documents and 200 safety justifications which had to be integrated together in a single document for aspects like the track, the power system, and so on – it was a paper mountain. In reality, at the back end of 2017, with the lateness of the trains and everything else, it was 60 per cent, not 90 per cent, complete.[3]

Moreover, says Wild, the supply chain was dysfunctional; but in this, the management faced a dilemma. One solution would have been simply to stop work, regroup and then complete the project. But the enormous supply chain could not survive that, and many of the contractors would simply have abandoned ship. Wild reckons that sorting out the mess with the suppliers took up to eighteen months: 'Nothing was coordinated, nothing was sequenced correctly and we should never have got into that situation.' In effect, all the risk of the project was now with Crossrail and that explains in large part why costs mounted significantly in the period after the abandonment of the December 2018 deadline.

Over the Christmas holiday break, Wild sat down with his team and decided on a strategy for how to complete the project. They were reluctant to set a date for the opening without undergoing a thorough assessment of where they were since the project was in such disarray. 'We decided to rebuild the programme. We sat down with the key people on Christmas Eve and decided to rebuild the programme in January because the main problem with Crossrail was that it was trying to be delivered all on day one.' Wild knew that opening one major station in central London was a big task, trying to complete nine simultaneously was near impossible. He said that when he was at London Underground, they reckoned it would take a year to eighteen months between finishing the installation of a station and opening it to the public. Originally the idea had been for Crossrail to hand over the whole railway in one go, but as Howard Smith, the chief operating officer, put it, 'no large project commissions like that. Having a staged railway is essential.'4 Most importantly, Wild wanted to decouple the railway – or route way as it is often called – from the stations. As he explained, 'each station might have 400 rooms, and 150 of them might be related to the routeway embedded in the stations like tentacles. This meant that the station contractors could never complete their work because they were always dependent on the route way.'5 Disentangling the stations from the route way was, therefore, top of Wild's 'to do' list, but it would take several months.

First he needed the right people to take over the project. Wild recognized that the number one priority was to get the right team in place to finish the job. While Wild is reluctant to criticize members of the previous top team, he is very clear that the wrong type of people were in charge. 'Systems integration', that all-too-nebulous but crucial term, requires a different set of skills from the civil engineering task of digging holes in the ground and putting in structures such as shafts, tunnels and stations. Not only did Wild have to get people in with the right skills, he had to replace or rehire many of those who had left.

Tony Meggs arrived as Crossrail chairman in January 2019 with the project in a parlous state and lacking a coherent leadership team after the series of departures in 2018. Ironically, his previous role had been as chairman of the Infrastructure and Projects Authority which, as mentioned in the previous chapter, had produced a series of favourable reports on the Crossrail scheme, giving it a reasonably confident green amber report, barely a year previously. Meggs says the first task for the project was to assess how much needed to be done: 'The notion that it was 90 per cent or so complete did not reflect what had been done. In some aspects, like testing, they had done 0 per cent. So the measure which is a construction measure, is useless in some respects, and what we had to understand was how to measure what still needed to be done *and that took us a year*' [my italics].

First, nor surprisingly, there was the political flak to deal with. Meggs recalls his first board meeting in January 2019:

It was a real meltdown. The first thing was dealing with anger from stakeholders, which was much of my role. The Department for Transport and TfL felt there must have been malfeasance and that people had lied to them. I had to convince them this was not the case. TfL in particular were angry because all of this was coming out of their budget and other projects had to be delayed and rescheduled as a result.

Meggs reports that he and Wild were very resistant to making commitments over a new opening date and a definite budget: 'TfL tried to make us agree to sign a piece of paper that the project would never cost more than a certain sum. To make matters more difficult, there were huge fights between TfL and the government and so dealing with the real anger was a big part of what I was doing.'[6]

While all these boardroom rows were taking place, construction work continued apace – with 7,000 people burning through £90m per month on thirty different sites.

There were two immediate tasks: improving the way that data was collected and changing the opening programme, so that not all the stations had to be delivered at the same time. Instead, there was to be a staged completion: once dynamic testing – which entails running trains to test out the system and iron out initial software bugs in the train control system – was completed, there were two further stages followed by a long period of trial running with trains operating to a timetable to flush out any further software bugs and test all the other systems with trains running in the tunnels. Gradually the number of trains was increased to move towards a full service. Then a third phase called 'trial operations' and lasting between three and six months, was envisaged.

Under pressure from both TfL and the Department, in April 2019 Wild relented and announced that Crossrail would be completed between October 2020 and March 2021, but that the Bond Street station would not be opened until later. It turned out to be a mistake: the new completion date would prove impossible to meet because – yet again – the extent of the work to be done had been underestimated.

As well as working out the completion programme, there was the major matter of the budget. Not only was every continued month of construction costing £90m, but significant increases in cost on some parts of the scheme were beginning to emerge. The stations, in particular, were proving far more expensive than originally envisaged. The National Audit office report on Crossrail published in May 2019 highlighted some remarkable cost increases at the stations. Nearly all were costing double the original target and several of the rises were far greater than that. Whitechapel was five times more expensive, having gone from £110m to £659m, and other stations with high increases included Woolwich (£70m to £234m) – this was surprising as Woolwich was a late addition to the scheme and there had been plenty of time for lessons to be learnt – and Farringdon (£239m to £634m). But it was not only expenditure on stations that was skyrocketing. The process of digging the tunnels, which had passed off relatively smoothly, still cost far more than initially estimated. The four main contracts had come in at almost double the expected amount, costing £2,120m rather than £1,170m.

Numerous explanations were given for these increases. But the key problem was poor project management and a failure to maintain control over the activity of the contractors, all exacerbated by the emphasis on meeting the December 2018 deadline. The NAO, for example, pointed out that 'during 2015 and 2016, some key contracts were moved from a target price to a cost reimbursement basis. This change meant that Crossrail Ltd removed the key incentive on contractors to minimise costs and took on the financial risk itself.'[7] The imperative of meeting the deadline, as noted in the previous chapter, also led to design changes taking place simultaneously with work in the field, which meant frequent re-planning of the programme. These late changes to the design of aspects of the construction and fit-out of both tunnels and stations led to conflicts between different contracts resulting in further delays and extra costs.

One particularly costly result of the permanent state of urgency born of the December 2018 deadline was that Crossrail began carrying out train and signalling system testing and construction activity in alternating time periods. This necessitated the demobilization of one set of workers in favour of the other and led not only to massive overtime payments but also to a situation in which certain individuals were unable to work but still had to be paid. Moreover, the need to get things done quickly meant that compensation claims by contractors were accepted with little scrutiny. Essentially the contractors were 'de-risked'; in other words, all the extra costs fell on Crossrail, not their contractors.

To be fair to Crossrail, some of the original budgets allocated to specific areas of the project were unrealistic, imposed on the project team by the arbitrary £1.6bn cut imposed by the then chancellor George Osborne in 2012 (described in Chapter 8), which had been made to fit in with a short-term political narrative. At that time the construction industry was still languishing in the post-crash doldrums and therefore contracts were accepted at unrealistic prices. In reality, the low contract prices and the commitment to meet Osborne's lower budget were always a fiction, especially given the well-known – one could almost say ubiquitous – tendency of major projects such as Crossrail to go over budget.

In response to these cost increases, which had an impact on the project both before and after the collapse of the 2018 deadline, bit by bit extra spending was granted, either through loans from the Department for Transport or by giving Transport for London the ability to borrow more. There was never any magic money tree for Crossrail. Every extra penny had to be fought for and negotiated with, at times openly hostile, ministers. The original 'funding envelope' of £14.8 billion had been increased just before the announcement of the delay to £17.6n; a further two sets of funding from Network Rail along with another rise in its budget in December 2020 resulted in the total reaching £18.8 billion. However, by mid-2021, Crossrail admitted it would top £19bn, since each extra month of work was costing in the order of £50m.

For the new Crossrail team, much of 2019 was a year of discovery. Every stone that was overturned seem to reveal yet more sources of cost and delay. There was constant pressure from the politicians to progress Crossrail but Wild, Meggs and the leading managers struggled to get the project under control. It was, in fact, not until the start of 2020 that Wild made the key appointment that was to make all the difference to progressing the project. Jim Crawford, like Wild a very experienced railwayman, was taken on as chief programme officer and it was not until he came on board that Wild considered that he had the right team to finish the job. Crawford had only just got his feet under the desk when he realized that there was no realistic prospect of opening in 2021, let alone 2020. Of course in March 2020 the Covid pandemic intervened and the 7,000-strong workforce were all sent home, with the exception of a group of 300 who were retained for safety reasons. However, Wild does not blame the pandemic for the subsequent delay as he recognizes that the schedule he had set out a year previously was not realistic. It took three months for work to restart fully but, interestingly, only 4,000 people were then on site.

As the workforce returned, on Crawford's recommendation, a different pattern of work was established. Rather than having four days of signal testing while no construction was taking place followed by three days of the opposite pattern with no trains being tested, Crawford opted for long blockades. Wild explained: 'we did much more intense signalling testing for seven days, then we would shut down the railway and focus entirely on construction for three to four weeks.'[8] There was even a period in the summer of 2020 when there was a blockade of the railway for six weeks, the aim being to allow an intense period of construction, with different teams working round the clock.

This change in the shift pattern to create long blockades, together with various other measures introduced by Crawford and the small team which came with him to Crossrail, led to a sharp improvement in productivity. Another innovation was the creation of the 'plateau' system, developed by Bombardier in Canada for its aircraft production process. The first 'plateau' was created for the signalling system. The idea was to bring together the various people engaged in a particular task – in this case, the signalling, train and platform screen contractors – to work in a collaborative way without recourse to the letter of their contracts. A second plateau system was set up for the station contracts where, again, numerous different contractors were brought together to resolve outstanding issues.

These improvements meant that instead of hitting on average around 30 per cent, average productivity, measured in terms of work done as a percentage of work scheduled, was doubled to at least 60 per cent and at times hit 90 per cent. The faster rate of progress meant that in August 2020, the board was able to announce that the project would open in the first half of 2022, but no firm date was given precisely because of the lessons learnt from having the earlier imposition of an overly rigid deadline.

The project was given added impetus in the summer of 2020 by the arrival of Andy Byford as Transport Commissioner at Transport for London. Byford, who had previously run transport networks in Toronto and New York, where he had been responsible for major investment projects, arrived just after Wild and his colleagues on the Crossrail board had decided it was impossible to open the line until early 2022 and only took the job on the basis that he would take personal charge of the Crossrail project. He says that he spends 60 per cent of his time on Crossrail, rather than leaving it to more junior managers at TfL. As with all newcomers to the scheme, Byford is reluctant to criticize anything that had occurred previously, but he clearly quickly realized that the project was still suffering from the trauma of the missed deadline. In order to ensure rapid progress and to reinforce the sense of urgency, Byford insisted that TfL – in other words him personally – have direct control over the project. He sought to have the Crossrail board abolished, a change which was duly effected within three months of his arrival. He also brought in from Bechtel two senior people whom he had worked with in the past to strengthen aspects of the leadership team which he felt were lacking in expertise. Keith Sibley was brought in as 'mobilization and improvement director' to support the transition from delivery to testing and subsequently to operation, while Mike Dunham – whom Wild lured out of retirement because of his specific expertise in completing stations – was put in charge of station delivery. Byford, on his arrival, had been struck by this aspect of the scheme: 'I was surprised by how much we still had to do to finish the stations. It shows that completing the project was not just about systems integration.'[9]

Byford took personal responsibility for relationships with the main contractors, notably Siemens and Alstom (who had taken over Bombardier and therefore become responsible for the trains): 'This is a showcase railway and the world is watching. I told the contractors that they had to give this job their number one priority above all their other contracts and they have responded to that.'[10] But he was swift to emphasize that there was no extra money as he was committed to working within the existing funding envelope.

Byford is rather bemused by the system that had been created to build Crossrail, notably the split between TfL and Crossrail over the procurement of the trains: 'It is strange that some of the stations are tube stations, others are the responsibility of MTR [the train operator] or Rail for London – it's bizarre, I would not have done that.'[11] He felt that this added complexity and blurred the lines of accountability.

The acceleration in the pace of work meant that in May 2021 the operation of trains in the tunnel could change from 'dynamic testing', which is intended only to uncover and fix bugs in the software system, to trial running, which involves running trains in the tunnels to demonstrate that the railway is safe and reliable as well as capable of meeting passenger capacity and performance requirements. The idea is to run the system under timetable conditions, gradually increasing the number of trains per hour, to flush out any remaining software bugs or other issues. It must be noted that the idea, as seemed to pertain in 2018, that this process could be undertaken from start to finish in around three months was never remotely realistic according to signalling experts. The third stage, trial operations, which involves testing the railway to ensure the safety and reliability of the railway for public use, finally started in November 2021 as this book was going to press. This included some large-scale volunteer exercises so that procedures such as rapid mass evacuation and train breakdowns can be played out in a fully operational environment with passengers. The opening date was ultimately contingent on dealing with the software bugs in the system as, in terms of construction and station fit-out, the Crossrail team was confident that this would be completed on time, with the possible exception of the still troublesome Bond Street station.

15.

A capital project

Many people, myself included, are sceptical of megaprojects. Such projects are all too often presented as the only possible solution when more modest schemes – or indeed other measures, such as encouraging people to adjust their behaviour or making societal changes – could solve the problem. There are legitimate criticisms to be levelled at Crossrail, and some awkward questions to ask. The original budget was evidently unrealistic, and has been replaced by a far more generous allowance, which used up all the large contingency and was in the end substantially exceeded – though an increase of around 30 per cent on the original budget, which had been artificially reduced, is not disastrous when set against some megaprojects across the world, notably the Boston Big Dig and the new Berlin Airport.

One could also argue that London has been very fortunate to acquire such a splendid new railway when so little is spent on transport in the regions in comparison with the capital. And it is reasonable to ask whether Crossrail goes to the right places, and whether its hugely expensive central tunnels are being used to best advantage. Most pertinently, it is a patently lopsided railway: even when it is fully operational half of its trains will be turned round at Paddington because no sensible destination has been found for them. There is no doubt that, in the future, this spare capacity will be used up, but that will involve further investment.

Andrew Adonis, who as transport secretary under Gordon Brown pushed through the decision to build High Speed 2, also worked hard to ensure that there would be space available at Old Oak Common for a Crossrail station that would connect with services on the high-speed line, enabling a rapid transfer between Heathrow and HS2. In the future, this interchange may well increase the number of Crossrail trains that go west beyond Paddington. A proposal has also been put forward that Crossrail trains serving Heathrow Terminal 5 could go through to Surrey, serving places such as Woking, Guildford and Basingstoke, but like many such projects, they so far remain on the drawing board.

There is, too, a rather large cloud hanging over the project. Already, in 2016, the number of rail travellers in London and the South East started falling, the first drop in passenger numbers for two decades (apart from a short blip after the 2008 banking crash). Crossrail was originally designed to accommodate 250 million passengers per year, a figure that should have been reached, according to the original plan, in 2020/21. Now, of course, this pattern of falling commuter numbers has been greatly amplified by Covid and expected changes in behaviour even after the pandemic is over. At the time of writing, in November 2021, passenger numbers on the Underground were hovering around 60 per cent of pre-Covid numbers. Moreover, there is a noticeable change in the pattern of use. At weekends figures at times reach 75 per cent, and this pattern is repeated on the national rail network where leisure travel has recovered far more quickly than commuting. There is little doubt that this will represent something of a permanent shift. Working from home is now accepted by nearly all employers and while it is likely that only a small number of people will never go into the office, it is also probable that the numbers of those who will, if their job allows, go into the office every day will be equally small. This will undoubtedly mean that Crossrail will struggle for many years to achieve the expected usage predicted when the go-ahead for construction was given. We have been here before. For several years after it opened in 1906, the Bakerloo Line, one of three Tube lines to open in quick succession, struggled to attract passengers and there were numerous press articles highlighting its failure.

Undoubtedly there will be similarly critical coverage once Crossrail – the Elizabeth Line, if we must – opens, but there will, too, be undoubted admiration for the extent, scale and sheer marvel of the new railway. There is no doubt that for all the failings listed in the past couple of chapters, Crossrail is an amazing scheme, grand in conception, executed efficiently and set to become as iconic for London as the red buses, the Tube or, indeed, Nelson's Column or the Houses of Parliament.

Despite the cost overrun and the delay, Crossrail has escaped rather lightly in the press. The focus of politicians and journalists on the Covid-19 pandemic since the late winter of 2020 may have drawn some critical sting. With a few exceptions, there have been few expressions of outrage at the waste of money in the tabloids. Even the considerable disruption caused by work at several central London sites, some of them in very sensitive areas such as the City and Mayfair, over much of a decade, has passed unnoticed or, at least, has attracted little adverse comment. Of course, the delay announced in August 2018 and the subsequent failure to open, together with the budget rises, has attracted negative coverage but it was relatively mild compared with that endured by other megaprojects when things went wrong. There has been no campaign of outrage in the tabloids and little detailed scrutiny – there should have been more – in the serious press. The project was greatly helped by the three series of programmes on its construction screened by the BBC. Remarkably, the third of these, aired in January 2019, attracted very little negative publicity, despite the fact that the line should have been opened by them.

Largely, Crossrail has been invisible while in plain sight. Despite those television programmes, few people understand the nature and, in particular, the scale of the project and the impact it will have on London and its transport system. I have lost count of the number of times I have had to explain the most basic details, even to people who have watched some of the TV programmes about the project. Perhaps Crossrail has got away so lightly because the project has been greatly undersold. MPs and ministers, and even TfL, have not made more of the fact that this is a brilliant public-sector project that has been built with a minimum of disruption and will be a major city-wide game-changer. Perhaps it is a classic case of the British with their love of understatement being embarrassed by their own achievements, together with the fact that the press, which is routinely hostile towards megaprojects, has made those involved in Crossrail understandably defensive. The City and the economists place great store by the economic benefits of the new railway. Their clever calculations, which I rather debunked earlier, suggest a significant increase (I won't give their figures credibility by citing them because, frankly, they are educated guesswork) in Britain's GDP thanks to Crossrail, but these numbers will not persuade Londoners of the value of the scheme. It is only when they venture down into the passageways and start using the trains that they will have an idea of the scale of the project and its impact on the capital.

As I keep having to explain to people fooled by the confusing decision to call it the Elizabeth Line, Crossrail is not a new Underground line: it is a national rail line using full-size electric trains with a remarkable 200-metre length. The stations are immense, several of them actually encompassing two existing stations. Farringdon reaches as far as Barbican, and Liverpool Street goes all the way to Moorgate. There is, quite simply, nothing like it. Londoners, in fact, may share the same experience as their Victorian forebears who were sceptical of the first Underground line completed in 1863 until they started using it and found it was one of the wonders of the age. Crossrail may well be viewed in the same light.

Moreover, the promises made about large sums of private money being invested in the project have not been realized. A National Audit Office (NAO) report published in 2014 found:

> The Department currently expects that one-third of the private sector funding it negotiated for Crossrail infrastructure will not actually be received. The Department negotiated agreements worth a total of £480 million, although it is not clear how the expected City of London Corporation

contribution was calculated. These contributions are now likely to total £320 million... This leaves a potential shortfall of £160 million which the Department will need to meet, from funds it had already set aside for the purpose.[1]

Indeed, one little observed aspect of the Crossrail scheme is that it is, in many ways, an old-fashioned project in that the funding of both the infrastructure and the trains was almost entirely on the government's account. It is a publicly funded project, sponsored by two state-owned organizations, Transport for London and the Department for Transport, and built by an organization with a very strong centralized ethos. Contrast this with the Channel Tunnel, which Mrs Thatcher mandated should be built on the basis of private-sector investment – though, in the event, it benefited from hidden government subsidies, such as British Rail's purchase of a set number of train paths. Gordon Brown's attempts to obtain private support for Crossrail largely failed, since investors do not like the long-term risks associated with megaprojects. It is cheaper for those drawing up such schemes to be honest about how much the private sector is prepared to put in, rather than trying to devise complex PFI-type arrangements that ultimately cost the state more and lead to delays while complex deals are negotiated.

Despite all the difficulties and the mistakes highlighted in the previous two chapters, a couple of aspects of the Crossrail methodology should offer a model for similar projects: innovation and legacy. From the outset, Crossrail placed innovation at the heart of the project's delivery, both to help address specific problems and to help future projects. Crossrail managers sought to build on practices established by major construction projects, including the London Olympics and Heathrow Terminal 5. They stimulated innovation through a programme called Innovate18, which was paid for by all the major contractors and match-funded by Crossrail. Everyone involved in the project was asked to submit ideas for improving safety or efficiency. Thousands of ideas were contributed and more than sixty trials funded. They ranged from drones for site surveys and 'smart' hardhats to various apps and new sensors to keep people and machinery separate. Probably even more than the specific ideas it was the culture of innovation the Crossrail team wanted to create that was important.

Another idea which permeated the entire organization was that of a legacy of learning, whereby the experience and insight gained through working on Crossrail would be shared with future projects. According to the Crossrail legacy website, 'documents and templates that have been used successfully on the Crossrail programme are provided to be "pinched with pride" by other projects'. This was a further advantage of the project being in the public sector. Private companies would have insisted that knowledge and experience become a source of profit. Crossrail, because it was funded by government, was able to pass on this information freely.

Terry Morgan had a particular commitment to apprenticeships. A product of a Welsh secondary modern school, Morgan went on to Birmingham University to study engineering and is acutely aware of the importance of education and training. He tells the story of how, in his previous job, as chief executive of Tube Lines, the public–private partnership company responsible for the maintenance and upgrade of the infrastructure of the Jubilee, Northern and Piccadilly lines, which was created by the Labour government in 2003 but collapsed seven years later when the maintenance was brought back in-house. He met a group of London Underground apprentices: 'I was shocked, all they could do was grunt.' So he set about ensuring that there would be a proper apprentice school attached to Tube Lines for signal engineers, who were then in short supply. 'It wasn't just about teaching them a trade, it was about communication skills, their expectations and so on.'

When Morgan arrived at Crossrail he was determined to do something similar, this time for tunnelling. His initial bid for government support was rejected, but he eventually managed to push through the funding of a permanent college dedicated to tunnelling (which was later expanded to include other disciplines): 'As a commercial proposition, it is highly questionable, but I got government funding for it and we managed to get a requirement in the contracts that for every £3m worth, the contractor would fund an apprenticeship.' Morgan made the point that he measured contractors' performances in the light of this commitment. If contractors failed to honour it, he 'would pull the chairman in and say you promised that you would do this'. It was one of the soft issues, such as community initiatives that Crossrail managed to include in contracts thanks to its sheer size. More than 1,000 apprentices have gone through the scheme. In 2017, Morgan said proudly, '30 per cent of our intake were women, and 40 per cent were NEETs [Not in Employment, Education or Training]'.[2] The college was later transferred to TfL and became a permanent feature.

Unfortunately, one damaging outcome of Crossrail's struggles to finish the project – together with the impact of Covid – is to make the prospect of a second line much less likely. Like the

Elizabeth Line, Crossrail 2 has been long in gestation, a period that stretches over half a century. The plan originally was for an Underground line between Chelsea and Hackney but this was later replaced by a more ambitious scheme. TfL considered whether it should be a metro scheme, with lots of stops and limited extensions at each end, or a regional railway, more like Thameslink, with few stations in the centre but extending far out into Hertfordshire in the northeast and Surrey in the southwest. And after a detailed assessment, the latter vision won out because, as with Crossrail 1, it would relieve pressure at mainline stations, notably Waterloo where growth has been particularly rapid. The line is also planned to go through the Upper Lea Valley, which was identified as an area where lots of housing could be built but only if there were better transport links.

The government was equivocal over the project even before Covid and consequently imposed a number of conditions before the scheme would be given the go-ahead. Notably, ministers insisted that local government would have to provide half the funding for a scheme costing in the region of £33bn and offer a clear strategy for financing the project. The politics of Crossrail 2 were already convoluted before the 2019 election. There was considerable hostility between the Conservative central government and Sadiq Khan, the Labour mayor, and this has only been exacerbated because the pandemic wrecked the economics of Transport for London, which has been forced to seek a succession of bail-outs to keep the trains and buses running. Now, given that in the 2019 general election, 'levelling up' was the key policy plank in order to help the North, the chances of getting political backing for another major scheme in the South East are slim. The argument that Crossrail, in fact, provided thousands of jobs in the north through the supply chain will simply not wash in the current climate. Therefore, as a condition for one of the bail-outs of Transport for London, the government forced TfL to put Crossrail 2 on ice, telling it to end any consultancy work, which rules out the scheme until there is a major shift in the political climate and a strong recovery from the post-pandemic reduction in passenger numbers. As with Crossrail, however, the route is to be safeguarded which means it will be possible to revive the project in the future.

The key issue will be how to raise the money. New sources of funding are essential if the scheme is to go ahead. Fares will be the biggest contributor, but TfL is acutely aware that it will have to go well beyond the methods used for Crossrail which, though quite innovative, still did not provide a sufficient proportion of the overall cost and failed to exploit some prospective sources of funding. Potential sources include: the developers who will benefit directly from the scheme; revenue from over-site development; a business rate supplement as with Crossrail 1 (though the fact that the existing extra charge does not run out till 2033 might prevent the imposition of a further one); payments Community Infrastructure Levy (which contributed greatly to Crossrail 1); and value extracted from land purchased for the railway. Ultimately, however, probably only road charging in the capital will provide sufficient funds.

All these funding options are hampered by the constraints currently imposed on the public sector as austerity remains the order of the day, and in the longer term by the fact that existing legislation prevents the use of many innovative ways of funding such schemes. Under the British system, the public sector creates a lot of value and wealth for the private sector, which largely gets a free ride. While Crossrail was partly funded by an increase in business rates, since this was applied across the capital there was no relationship between the specific uplift in value created by Crossrail and the extra taxes paid. The official report on the impact of Crossrail on property values makes it clear that much more could have been done to use some of the increased value resulting from the new line to help fund it.

For the time being, therefore, London's rail travellers will have to enjoy the Elizabeth Line. Mark Wild has insisted that it will only open if he is certain it will provide a good service as the type of first-day experience which occurred with Heathrow Terminal 5 caused long-lasting – though not permanent – damage. Wild wants no such mishaps with Crossrail. Provided this is adhered to, Crossrail will soon win over Londoners and the delays and cost overruns will quickly be forgotten. Who ever mentions that the Jubilee Line was scandalously over budget or that the British Library was a disaster? Even the most cynical reporter will be unwilling to pour scorn on an engineering marvel that will quickly become a vital part of London's transport system. People will try to use sections of it whenever they can just because the experience will be so much more pleasant than straphanging on the Central Line. This might be the only hope for Crossrail 2. People using Crossrail will be so struck by the efficiency and comfort of a modern urban railway that they will demand more. It might take a bit of time, but hopefully the clamour will eventually become great enough.

Acknowledgements

Enormous thanks go to the Crossrail team who were all incredibly open, despite their concerns about the fact that I had been critical of the scheme in the past. Terry Morgan, the chairman, was particularly ready to give me his time and opened all the doors for me. Peter MacLennan, who had the onerous job of liaising with me, was extremely thorough and deserves special thanks both for answering my (precisely) fifty queries in double quick time, and for organizing countless visits. Others at Crossrail who helped and gave me generously of their time included, and forgive me for any omissions: Simon Bennett, Chris Binns, Julian Robinson, Bill Tucker, Andrew Wolstenholme and Simon Wright. Thanks, also, to all those, rather too many to mention, on site who I met and who showed me round stations, tunnels, depots and control centres.

I would also like to thank the many others who gave me their time for interviews or helped in other ways, including numerous people who had previously been involved in the scheme: Andrew Adonis, Keith Berryman, Andrew Bosi, Rupert Brennan Brown, Liam Browne, Michael Cassidy, Michele Dix, Don Heath,

Peter Lewis, Richard Malins, Richard Meads, Michael Schabas, Howard Smith, Jim Steer, Chris Stokes, David Warren and Jon Willis.

Bernard Gambrill, who worked for Crossrail, and who has helped me on several other books, again read the manuscript and made many useful comments. And thanks to Richard Milbank who is the most diligent editor I have worked with (and that is saying something); and to Toby Mundy, my excellent new agent, who seems to manage to ensure I continue to be gainfully employed.

As ever, my wife, Deborah Maby has been ridiculously supportive and put up with rather more stress than I normally experience as the deadline loomed sooner than I had hoped. And my kids are no longer a distraction as they have all left the roost but instead beat me at tennis and running and inspire me to battle on.

The manuscript has not been shown to anyone at Crossrail and therefore any errors that have slipped through are, of course, entirely my responsibility.

Notes

1. The First Crossrail

1 Jerry White, *London in the 19th Century*, Vintage, 2008, p. 76.
2 Wayne Asher, *A Very Political Railway: The Rescue of the North London Line*, Ian Allan, 2014, p. 10.
3 Quoted in Barbara Denny, *Notting Hill and Holland Park Past*, Historical Publications, 1993, p. 31.
4 *The Railway Times*, 19 July 1890.
5 *Financial Times*, 28 June 1900.

2. The Crossrail Concept

1 *The British and Foreign Railway Review*, 'Regent's Canal Railway', 1845, p. 306.
2 Public Record Office, CAB Cabinet Office, 134/915.
3 A scanned version of this very old-fashioned and long-forgotten pamphlet is available at http://www.dragondark.co.uk/lr/crossrail-1980londonreconnections.pdf
4 Interview with the author.
5 Interview with author.
6 Department of Transport, *Central London Rail Study*, 1989, p. 3.
7 Ibid.
8 Ibid., p. 4.
9 Interview with author.
10 Michael Schabas, *The Railway Metropolis: How Planners, Politicians and Developers Shaped Modern London*, Institution of Civil Engineers, 2017, p. 96.

3. Megaprojects and Mega-businesses

1 Michael Schabas, *The Railway Metropolis: How Planners, Politicians and Developers Shaped Modern London*, Institution of Civil Engineers, 2017, p. 131.

2   Interview with author.
3   Central London Rail Study, A joint study by The Department of Transport, British Rail Network SouthEast, London Regional Transport and London Underground Ltd, January 1989, p. 19.
4   Bent Flyvbjerg, with Nils Bruzelius and Werner Rotherngatter, *Megaprojects and Risk: An Anatomy of Ambition*, Cambridge University Press, 2003, p. 44.
5   Ibid.
6   Ibid., p. 45.
7   Quoted in ibid., p. 48.
8   Omega Centre, *Megaprojects Executive Summary: Lessons for decision-makers: an analysis of selected international large-scale transport infrastructure projects*, UCL, 2012, p. 16.
9   Ibid., p. 20.
10 Omega Centre, *Megaprojects Executive Summary*, p. 36.
11 Schabas, *The Railway Metropolis*, p. 52.
12 Ibid., p. 85.
13 Ibid., p. 51.
14 Ibid., p. 55.
15 Quoted in Jon Willis, *Planning the Jubilee Line Extension*, London Transport, 1997, p. 38.
16 Schabas, *The Railway Metropolis*, p. 106.
17 Jon Willis, *Planning the Jubilee Line Extension*, London Transport, 1997, p. 43.
18 House of Commons Transport Committee, London's Public Transport Capital Investment Requirements, 1992/3 session, third report, HC 754, p. 29.

4. Saved But Shelved

1   Michael Schabas, *The Railway Metropolis: How Planners, Politicians and Developers Shaped Modern London*, Institution of Civil Engineers, 2017, p. 198.
2   Ibid., p. 199.
3   Quoted in Steve John, *The Persuaders: When Lobbyists Matter*, Palgrave Macmillan, 2002, p. 111.
4   Ibid.
5   Schabas, *The Railway Metropolis*, p. 201.
6   All these quotes come from interviews with the author.
7   Interview with author.
8   Interview with author.
9   Christian Wolmar, 'Back to the drawing board: the MPs who nixed

Crossrail have been pilloried but they were right', *Independent London*, 12 May 1994.

10 John, *The Persuaders*, p. 108.

11 Quoted in John, *The Persuaders*, p. 109.

12 Bovis et al., CrossRail Effectiveness, Department of Transport, December 1993, p. 8.

13 Interview with author.

14 Christian Wolmar, 'London's £3bn Crossrail at risk as Bill is shelved', *Independent*, 11 May 1994.

15 Leo Walters, Evidence to the Opposed Bill Committee, Crossrail, 8 May 2008.

16 Quoted in John, *The Persuaders*, p. 126.

17 Interview with author.

18 Interview with author.

19 Interview with author.

20 Christian Wolmar, 'Back to the drawing board, the MPs who nixed Crossrail have been pilloried but they were right', *Independent London*, 12 May 1994.

21 Interview with author.

22 Christian Wolmar, 'Back to the drawing board, the MPs who nixed Crossrail have been pilloried but they were right', *Independent London*, 12 May 1994.

23 Hansard, House of Commons debates, June 8 1993, Vol. 226, cc201–47.

24 Quoted in John, *The Persuaders*, p. 126

5. Crossrail Revived

1 Christian Wolmar, 'Back to the drawing board, the MPs who nixed Crossrail have been pilloried but they were right', *Independent London*, 12 May 1994.

2 Ibid.

3 Interview with author.

4 Interview with author.

5 Strategic Rail Authority, *London East-West Study*, 2000, p. 13.

6 Ibid., p. 16.

7 Ibid., p. 8.

8 Interview with author.

9 Crossrail Business Case, Cross London Rail Links, Transport for London and Strategic Rail Authority, 2003, p. 4.

10 Ibid., p. 7.

11 Ibid., p 8.
12 Ibid., p. 7.
13 The explanation for the need for the Review of the Crossrail Business Case, Department for Transport, July 2004, p. 13.
14 Christian Wolmar, *Down the Tube: The Battle for London's Underground*, Aurum Press, 2002, p. 92.
15 Review of the Crossrail Business Case, Department for Transport, July 2004, p. 13.
16 Ibid. p. 108.
17 Ibid., p. 104.
18 Interview with author.
19 Review of the Crossrail Business Case, Department for Transport, July 2004, p. 105.
20 Interview with author.
21 Email to author.
22 Michael Schabas, *The Railway Metropolis: How Planners, Politicians and Developers Shaped Modern London*, Institution of Civil Engineers, 2017, p. 225.
23 *Financial Times*, 26 May 2004.
24 House of Commons, 20 July 2004, col. 159.
25 Ibid., col. 25.
26 *Daily Telegraph*, 1 February 2004.

6. Seeing off the Naysayers

1   Review of the Crossrail Business Case, Department for Transport, July 2004, p. 57.
2   Ibid., p. 59.
3   Ibid., p. 62.
4   Ibid., p. 63.
5   Review of the Crossrail Business Case, Department for Transport, July 2004, p. 97.
6   Ibid.
7   Ibid.
8   House of Commons Debates, 7 April 2005.
9   Ibid.
10 Michael Schabas, *The Railway Metropolis: How Planners, Politicians and Developers Shaped Modern London*, Institution of Civil Engineers, 2017, p. 202.
11 Ibid., p. 203.
12 Ibid., p. 199.

13 Interview with author.

14 Schabas, *The Railway Metropolis*, p. 204.

15 *Non technical summary of the Crossrail Environmental Statement*, p. 54.

16 *Crossrail Environmental Statement*, Vol. 3, p. 30.

17 Ibid., p. 31.

18 This and subsequent two quotes from *Crossrail Environmental Statement*, Vol. 3, p. 31.

19 This and subsequent quotes in this section from *Crossrail Environmental Statement*, Vol. 3, p. 32.

20 Ibid.

21 https://learninglegacy.crossrail.co.uk/learning-legacy-themes/environment/biodiversity/

22 Department for Transport, *Crossrail race equality impact assessment, first report of the full assessment*, p 19.

23 Institution of Civil Engineers, *Crossrail project: programme managing the Elizabeth Line, London*, November 2017, p. 13.

24 Ibid., p. 14.

7. Money, Money, Money

1 Department for Transport and Transport for London, *Heads of terms in relation to the Crossrail Project* http://webarchive.nationalarchives.gov.uk/20080809024303/http:/www.dft.gov.uk/162259/165234/302038/headsofterms.pdf

2 National Audit Office, HC 965 Session, 2013/14, 24 January 2014, p. 21.

3 Michael Schabas, *The Railway Metropolis: How Planners, Politicians and Developers Shaped Modern London*, Institution of Civil Engineers, 2017, p. 226.

4 Ibid.

5 *Rail* magazine, 14 January 2009.

6 Interview with author.

7 This and subsequent quotes are all from interview with author.

8 Interview with author.

9 Interview with author.

10 Interview with author.

8. A Daunting Task

1   London Assembly Transport Committee, *Light at the End of the Tunnel, the Construction of Crossrail*, February 2010, p. 18.
2   National Audit Office, HC 965 Session, 2013/14, 24 January 2014, p. 22.
3   Ibid.
4   All Terry Morgan quotes are from interviews with author.
5   Interview with author.
6   All the quotes from Terry Morgan are from interviews with author.
7   All Bill Tucker quotes are from interviews with author.
8   *Rail Technology Magazine*, Feb/Mar 2014.
9   Institution of Civil Engineers, *Crossrail Project: Designing and Constructing the Elizabeth Line*, London, May 2017, p. 8.

9. Digging Under London

1   Institution of Civil Engineers, *Crossrail Project: Designing and Constructing the Elizabeth Line*, London, May 2017, p. 24.
2   http://news.bbc.co.uk/1/hi/england/london/8574619.stm
3   Gillian Tindall, *The Tunnel Through Time: A New Route for an Old London Journey*, Chatto & Windus, 2016, p. 15.
4   Institution of Civil Engineers, *Crossrail Project*, p. 23.
5   Ibid., p. 24.
6   Hugh Pearman, *Platform for Design*, Crossrail, 2016, p. 10.
7   Interview with author.
8   Institution of Civil Engineers, *Crossrail Project*, p. 48.
9   Email to author.
10 Institution of Civil Engineers, *Crossrail Project*, p. 38.
11 Ibid., p. 37.
12 Crossrail website: https://www.youtube.com/watch?time_continue=150&v=vBkb1dS9QB4
13 *Rail Technology Magazine*, 23 July 2017.
14 Interview with author.
15 Tindall, *The Tunnel Through Time*, p. 14.
16 Email to author.

17 Institution of Civil Engineers, *Crossrail Project*, p. 38.
18 Ibid.

10. Stations for the Future

1   Interview with author.
2   Interview with author.
3   Interview with author.
4   Hugh Pearman, *Platform for Design*, Crossrail 2016, p. 11.
5   Ibid.
6   Ibid., p. 25.
7   Ibid., p. 26.
8   https://www.londonreconnections.com/2015/crossrail-progress-paddington/
9   Interview with author.
10 Pearman, *Platform for Design*, p. 93.
11 Interview with author.
12 Interview with author.
13 Interview with author.
14 *Railconnect*, May 2012, p. 23.
15 Institution of Civil Engineers, *Crossrail Project: Designing and Constructing the Elizabeth Line,* London, May 2017, p. 57.
16 Ibid.
17 Ibid., p. 62.

11. Trains and Tunnels

1   Interview with author.
2   Interview with author.
3   *Rail* magazine, 6–19 June 2018, p. 54.
4   Interview with author.
5   *Independent*, 31 August 2011.
6   Ibid.
7   Interview with author.
8   Email to author.

12. The Finishing Touches

1   Interview with author.
2   Interview with author.
3   Interview with author.

13. And Another One?

1   Foresight, July 2016 https://assets.kpmg/content/dam/kpmg/pdf/2016/07/foresight-44-crossrail-project.pdf
2   Jacobs, Project Status Report 106, Period 09 FY2017-18 12, November 2017–09 December 2017, No. B2111500/106/1.19 10, January 2018.
3   Jacobs, Project Status Report 111 Period 1 FY2018-19 01 April 2018 – 28 April 2018 Document No. B2111500/111/1.18 24 May 2018.
4   Interview with author.
5   House of Commons Committee of Public Accounts Crossrail: progress review Ninety-Second Report of Session 2017–19, HC 2004, April 3rd 2019, p 12.
6   House of Commons Committee of Public Accounts Crossrail: progress review Ninety-Second Report of Session 2017–19, HC 2004, 3 April 2019, p. 12.
7   https://www.londonreconnections.com/2019/crossrail-the-dangerous-sound-of-silence/
8   Crossrail Press Release, 8 March 2018.
9   All these quotes, interview with author.
10 Interview with author.
11 Interview with author.
12 Interview with author.
13 Interview with author.
14 Interview with author.
15 Interview with author.
16 https://www.londonreconnections.com/2019/crossrail-the-dangerous-sound-of-silence/
17 Interview with author.
18 Interview with author.
19 Interview with author.
20 Interview with author.
21 National Audit Office, Completing Crossrail, HC 2106 session 2017–2019, 3 May 2019.
22 Interview with author.
23 Interview with author.

14. Rescue

1   https://spectrum.ieee.org/londons-crossrail-is-a-21-billion-test-of-virtual-modeling#toggle-gdpr

2   https://www.londonreconnections.com/2018/crossrail-breaking-down-the-crisis/
3   Interview with author.
4   Interview with author.
5   Interview with author.
6   Both quotes, interview with author.
7   National Audit office, HC 2106, 2017–2019, 3 May 2019, p 8.
8   Interview with author.
9   Interview with author.
10 Interview with author.
11 Interview with author.

15 A Capital Project

1   National Audit office, Crossrail, HC965, January 2014, p. 25.
2   Interview with author.

Index